THE GREAT AGES OF DISCOVERY

STEPHEN J. PYNE

THE GREAT AGES
OF DISCOVERY

How Western Civilization Learned About a Wider World

THE UNIVERSITY OF
ARIZONA PRESS

TUCSON

The University of Arizona Press
www.uapress.arizona.edu

ISBN-13: 978-0-8165-4111-9 (hardcover)

Cover design by Leigh McDonald
Cover images: *Columbus in India primo apellens, magnis excipitur muneribus ab Incolis. Columbus landing in Hispaniola*, engraving by Johann Theodor de Bry (1561–1623), National Maritime Museum, Greenwich, London [top]; The crew of the "Fram" in the Bay of Whales, from *The South pole; an account of the Norwegian Antarctic expedition in the "Fram,"* *1910–1912*, by Roald Amundsen [center]; Cygnus space frieghter, NASA [bottom]
Designed and typeset by Leigh McDonald in Minion Pro 10.25/15 and Telmoss WF [display]

Library of Congress Cataloging-in-Publication Data
Names: Pyne, Stephen J., 1949– author.
Title: The great ages of discovery : how western civilization learned about a wider world / Stephen J. Pyne.
Description: Tucson : University of Arizona Press, 2021. | Includes bibliographical references and index.
Identifiers: LCCN 2020018149 | ISBN 9780816541119 (hardcover)
Subjects: LCSH: Discoveries in geography—History. | Voyages and travels—History.
Classification: LCC G80 .P96 2021 | DDC 910.9—dc23
LC record available at https://lccn.loc.gov/2020018149

Printed in the United States of America
♾ This paper meets the requirements of ANSI/NISO Z39.48-1992 (Permanence of Paper).

To Sonja
a polestar, as ever
to Lydia and Molly
who tolerated my errant wanderings over the years
and
to William H. Goetzmann
who gave me a scholarly sextant by which to navigate

Contents

BOOK III. MISSIONS OF DISCOVERY

THE GREAT AGES OF DISCOVERY

FRANCISCI
DE VERULAMIO/
Summi Angliæ
CANCELLARIJ/
Instauratio
magna.

Multi pertransibunt & augebitur scientia.

LONDINI
Apud Joannem Billium
Typographum
Regium.

Anno   1620

Francis Bacon, *The Great Instauration* (1620), frontispiece.
Sailing beyond the Pillars of Hercules.

Prologue

Europe on the Edge

And most important of all was the coast of Portugal.[1]

—CARL SAUER, *NORTHERN MISTS*

When the 14th century ended, Europe occupied a marginal part of the known world, and a far smaller part of the world beyond. Even to the heartland of European civilization, then still clustered around the Mediterranean Sea, "Atlantic Europe was the farthest back country," as Carl Sauer observes, full of barbarians and the "shores of a difficult and repellent sea, *mare tenebrosum*," a Sea of Darkness.[2]

Geographically, Europe claimed the splintering western flank of the great Eurasian landmass. It appeared as a shrinking tangle of peninsulas and islands, like the fraying end of a rope, that dissolved into the Atlantic Ocean. Economically, it remained feudal and had few goods or services the rest of the world wanted other than bullion. It always teetered on famine, had no significant manufacturing, and found itself on the deficit side of commerce. Demographically, it ended the century with half the people with which it had begun. The Black Death, ravaging town and country from 1347 to 1351, had killed 30–60 percent of its population. Militarily, it endured such internal disasters as the Hundred Years' War, and it was retreating from the advances eastward it had made during the Crusades. In 1396 it suffered a crushing defeat at Nicopolis from the expanding Ottoman Turks. In 1379 even the patina of unity provided by Christendom was rent when two popes were elected, one in Avignon, and one in Rome, announcing the Great Schism; thirty years later, a third pope was elected. Culturally, its first renaissance of learning that had flourished in the 12th and 13th centuries had sunk into a scholasticism that had too many commentators and not enough texts to comment on. A rebirth of classical learning was underway in Italy, but the scientific revolution was a good three centuries in the future. What Europe knew about a wider world it gleaned from ancient texts and a handful of travelers and traders who had reconnected with the east during the Pax Mongolica. In many ways Europa knew less about the world than it had a millennium before.

What the rest of the world knew about Europe is unclear. It was probably not much, and probably not of much concern except to immediate neighbors. Europe was not a destination for scholars, missionaries, travelers, artists, or traders. It was hemmed in on the east by hostile, more powerful peoples, and on the west by the horizon-eclipsing Sea of Darkness.

Yet, over the course of the coming century, Europe would take to the world ocean, and by the mid-16th century it would connect with all the populated continents, establish commercial and military empires that spanned both the New and Old Worlds, and still others yet more exotic, and inscribe a map of the planet that all peoples would come to accept. Spearheaded by geographic exploration, it moved from the margins of the world to something like its center.

In retrospect, it is obvious that Europe's expansion had to be seaborne. But it was not obvious at the time. Europe had little that was special by way of blue-ocean technology. Voyages westward had repeatedly stalled or been lost. Nor was it alone in having a maritime tradition, or in sailing beyond its home range. It was but one of a suite of commercial societies built on shipping.

The major civilizations centered around Eurasia, or rather its margins. China dominated the Pacific Coast; India, the Indian Ocean, extending into Indonesia; and the Islamic ecumene, with its southern rims spilling across northern Africa and penetrating into the South China Sea. These were joined by trade routes that traced back to antiquity. In the Americas a succession of civilizations flourished in Mesoamerica, with special concentration in the Valley of Mexico, and in the highlands of present-day Peru. Apart from these primary hearths, there were many others of regional or subcontinental significance. Africa had several—the Zimbabwe and the empire of Mono-motapa in the south, Ghana in the west, full of gold and slaves, and Ethiopia in the east, better known to Europe as the Kingdom of Prester John, a Christian principality with which Europe might forge an alliance against Islam.

Each people naturally considered themselves as the core, and created maps and narratives that placed their homeland at the center of the world and their story at the core of the world's narrative. Curiously, Europe was an exception. Its conversion to Christianity put the center of its *mappa mundi* in the Holy Land. The Roman Empire that had defined Europe for centuries was split, north and south, east and west. Islam claimed the south; Christendom, the north; they fought in Spain, the Balkans, and throughout the Mediterranean Sea. The memory of Rome also left Latin Europe on the fringes or loosely linked at hinges such as Venice. It was marginal to world trade, military might, and intellectual achievement, still harking back to an imagined golden age when philosophers spoke Greek and statesmen Latin.

Most of these civilizations had a maritime component: through the sea they could reach out and in turn be reached. (The exceptions were the Aztec and Inca empires, which remained landlocked.) But there were maritime cultures that were regionally important, not all of which map onto this collage of civilizations. The Caribbean islands were successfully colonized from the mainland by boat, isle by isle like a stone walkway, the Lesser Antilles to the Greater, and remained in easy contact among themselves. The Indian Ocean sustained a widespread maritime tradition, powered by the seasonal monsoon winds that blew north in the summer and south in the winter. With India as midpoint, monsoon-driven trade extended from Africa and the Red Sea to the Spice Isles and present-day Indonesia. This realm overlapped with the South China Sea that brought China and Southeast Asia into contact.

Well within the monsoonal winds, the region was a kind of triple point. It overlapped with the great Austronesian diaspora that spread from offshore China through the South China Sea in two grand sweeps. One contoured around the Indian Ocean and ended in Madagascar (landing around 100–500 CE). The other island-hopped across the Pacific, well beyond the monsoonal regime. It reached Samoa around 1000 BCE, paused before advancing to Tahiti (700 BCE) and the Marquesas (300 BCE). Then it paused again before a starburst of voyaging took it north to Hawai'i (800 CE), southeast to Easter Island (400 CE) and southwest to New Zealand (900–950 CE). Voyagers reached Easter Island roughly when their Malagasy cousins did Madagascar. They colonized New Zealand around the time the Norse did Greenland.

If the Polynesians sailed farther than anyone else, the Ming dynasty had sailed more elaborately in order to "display the wealth and power" of China. In seven massive voyages under Zheng He from 1405 to 1433, Chinese fleets journeyed throughout Indonesia, through the Strait of Malacca, to India, to Sri Lanka, to Persia in the Gulf of Hormuz, to the Red Sea, to the east coast of Africa, and even perhaps to the Cape of Good Hope. The 1405 voyage reportedly had 62 ships and 27,800 men; the scale is an order of magnitude greater than what Europeans could muster. In odd counterpoint, China was voyaging along Africa's east coast about the time Portugal was probing piecemeal along its west coast. Had China continued and Portugal quickened its pace, they might have met while doubling the cape.[3]

Each maritime culture had undergone pulses of expansion and contraction. Greater Polynesia's had waxed and waned over the course of centuries. It took 300 years to go from Samoa to Tahiti, and another 300–400 to reach the Marquesas. From there 1,100 years passed before Hawai'i, and 1,200 before New Zealand. China's Great Voyages exploded suddenly on a colossal scale, and imploded just as rapidly 28 years later. Europe was no exception. Because of its crenulated coastline, most European principalities or nations had access to the sea. Most of its great migrations, however, had come overland from the east in repeated surges. Its maritime adventures had come and gone. In Roman times Europe had reached beyond the Mediterranean to the Canary Islands (known as the Fortunate Isles) and had traded with India. During the 8th–10th centuries Norse expansion had colonized Iceland and Greenland, then retreated. European history was full of false dawns that left populations isolated and sites withered into ruins. In the 14th century Europe was struggling to regain what it had once known and held, and then forgotten and lost.

Only the rare and reckless speculator would have bet that Europe would be the civilization that united the others, that connected all the dots and drew the reigning mappa mundi, much less that the margins of marginal Europe would be the cutting edge of the future.

Yet Europe had assets whose significance only became apparent later. Perhaps critically, the endless squabbling among Europeans that made it hard to consolidate the small continent under a single rule and that made it difficult to muster a large-scale or systematic effort at geographic discovery, also created rivalries that ensured that, unlike Ming China, competition could not be exiled by decree.

Early modern Europe and exploration coevolved. Europe had a sufficient culture of literacy and enough centers of learning that discovery could be studied and preserved, and the printing press meant new accounts and revised maps could be disseminated widely. Johannes Gutenberg opened his press in 1439, 19 years after the discovery of Madeira, and 29 after the recovery of Ptolemy's *Geographia*. Equally, Europe was hybridizing ship designs that yielded a flotilla of vessels capable both of close coasting and of long-distance transport. It had serious military technology and traditions, with notable advances in naval ordnance that allowed ships to become floating batteries, suitable for attacking other ships, not simply platforms for soldiers to fight from. It had maritime traditions that rose from long folk traditions. With a coastline longer, and proportionately greater by far, than that of any other continent, Europe could look to the sea as a natural corridor for expansion. It had a tradition of expanding and a literature of romances that celebrated adventure and quests, most recently through the Crusades. Not least it had a hunger, a vigor, a curiosity that could morph into a yearning that could mutate into ruthlessness, an ambition that bordered on lust, and a festering restlessness that could be channeled to purposeful discovery.

Other civilizations possessed some or most of these same features. But only Europe managed to institutionalize them in ways that lurched and rolled along for centuries. Geographic discovery became a distinct practice, not something that just spilled out like travelers' tales from random trade, pilgrimage, raids, and wandering. It rose and fell with the greater tides of its sustaining culture, but when it surged it was a full-immersion engagement such that over the centuries it became difficult for thoughtful observers to imagine a Western civilization without it.

What happened over the next 600 years was in some ways foreshadowed, or at least had its pieces fashioned, in the Europe of 1400.

Begin with geography. Europe's physical geography resembles a welter of small seas and straits. Two are internal, the Baltic and the Mediterranean. But thanks to islands, each could be subdivided into smaller seas that straits and short land passages could connect to other seas or rivers that led to external seas, the Black, the Red, the North, the West. Those passages were portals, at once points of bottlenecks and points of egress. Most sailing was coasting, most navigation was piloting, most maps traced the littoral. Ships passed from sea to sea, hauling cargo, troops, and pilgrims from island to island, shoreline port to port. Ships had sails but mostly moved by oars. The Great Voyages across the world ocean relied on similar islands as fortresses cum factories and ports, and searched feverishly for straits that could connect ocean to ocean.

But terrain is not fate. What mattered was how Europe's peoples transformed that geography into history. The global maritime empires that swelled in the wake of the Great Voyages followed precedents based on the competition over trade and geopolitics that raged among Europe's internal seas, especially the Mediterranean. Those contests dated back to ancient times. The most famous was the rivalry between Rome and Carthage; the most immediate, the new colonization of Mediterranean islands by Catalonia and the formidable competition between Genoa and Venice within a larger contest between Latin Christendom and the Ottoman Turks. Using a similar playbook, relying on some of the same players, Spain and Portugal projected that legacy beyond the Pillars of Hercules out to the Ocean Sea.

Both Venice and Genoa were city-states, both committed to the sea for their livelihood, both were prodded by that foray of Europe eastwards known as the Crusades, and both competed, with lethal intent, for the same commerce. They were famously different in character. Genoa was a loose association of commercial freebooters, little subject to any governing authority, opportunistic and pugnacious, willing to push into new territory but as likely to turn to piracy—condottieri of the seas. They traded whatever was at hand, but their most lucrative traffic was slaves. Venice was more closely ruled with a governing council overseen by a doge, had a state navy, was inclined to take over existing trade routes rather than probe into the unknown, and orchestrated its overseas possessions into a roughly coherent whole. It specialized in high-end luxury goods, of which spices were the best known. In a celebrated sketch Franco Sacchetti of Florence compared the Genoese to donkeys and the Venetians to pigs. When donkeys are together "and one of them is thrashed with a stick, all scatter"; when pigs gather and one is struck, "all draw close and run unto him who hits it." The "antithesis" between the Genoese and the Venetians, in the words of G. V. Scammell, was "total."

Moreover, should either falter, there were plenty of others like Pisa and Catalonia ready to step into the breach.[4]

From the onset of the 13th century to the end of 15th the two self-styled republics fought each other for supremacy, and each, separately, fought the new great power to the east, the Ottoman Turks, who eventually amassed a navy as great as its army and overran their posts as it did other outliers of a European presence in Asia. Each had its critical island sentinel—for Genoa, Chios in the Aegean; for Venice, Crete, astride the Aegean and the eastern Mediterranean approaches to the Levant. From posts on the Sea of Azov they connected with the overland silk road, and from contacts with the Red Sea, they learned of the maritime routes to India, the Spice Isles, and China. Two Venetian travelers, Marco Polo (1254–1324) and Niccolò de' Conti (1395–1469), wrote dazzling accounts of their experiences in the Indies and Cathay and of the wealth those distant lands exhibited. Italian trading factories ranged from North Africa to northern Europe.

Europe knew well what the east offered: the problem was reaching it when travel meant passing through lands controlled by Arabs and Turks hostile to interlopers, especially when they were also long-standing religious rivals. Even during the Crusades Europeans had sought an oceanic workaround; in 1291 the Vivaldi brothers sailed galleys in a probable effort to circumnavigate Africa (they disappeared without a trace). For a while the Mongolian empire had opened overland transit and invited travelers, but its collapse left the east barred. Genoese mariners turned to the Atlantic. By the 1330s mariners were probing into the Atlantic; by 1347, they had reconnected with the Canary Islands, once known in Roman times, but since forgotten. The isles critical to Europe's Mediterranean empires now relocated into the Atlantic.

Their rivalry was a template for what was to follow as late medieval times segued into a Renaissance and quarrels among Europeans were flung about the globe. The critical components clicked into alignment, like tumblers in a lock. Competitors waiting to seize the initiative should one of the principals falter; fights over distant trading colonies; commerce backed by military force; an interest in trade, not settlement; the role of islands; models for governance of overseas factories and possessions; exemplars for exploring new lands through state sponsorship or through state-sanctioned private companies; the practice of chronicling voyages and of recording their discoveries in maps—all this would reappear on much vaster scales in the centuries after Genoese and Venetian galleys ceased to ply the waves.

Nor was their role merely symbolic and analogic. The Genoese especially were active in Iberia; they helped finance expeditions; they served as pilots, advisors, and mentors; they worked with Portugal and Spain to transform newly discovered lands such as Madeira and the Canaries into working sugar plantations, staffed with slaves. Chios and Crete proposed differing models for settlement and governance. The pioneer pilot to the Americas, Christopher Columbus, was Genoese. The first English explorer to North America, John Cabot, was Venetian; so was the great chronicler of Ferdinand Magellan's Armada de Molucca, Antonio Pigafetta. For a century after the voyages that blazed routes to the Americas and India, the two city-states continued to be active, directly or indirectly trying to reassert their primacy until exhaustion and a shift in Europe's geopolitical center northward and westward left them as tidal wash on the littoral of exploration history.

In 14th-century Italy a ferment emerged that transitioned between medieval and modern times. The movement spread, gradually percolating throughout Latin Christendom. The Renaissance, as it became known, birthed the discovery of two new worlds, one of learning and one of geography. As its name suggests, it also found lands known in past times and lost, or legendary, and now sought, like a missing manuscript from Aristotle.

The project was a self-conscious recovery, a literal "rebirth." Europe began reconnecting with ancient learning in the form of rediscovered texts. The catalytic personality in what became known as humanism may have been Petrarch (1304–1374). In 1336, in rough imitation of a passage from Livy, and carrying with him a copy of St. Augustine's *Confessions*, he climbed Mont Ventoux to inspect the view and meditate. In 1345 he discovered a cache of Cicero's letters, a galvanic moment for the humanist movement. Those milestone events correlate nicely with the renewal of geographic discovery: a map from 1339 includes a record of discovery of the Canaries by the Genoese mariner Lanzarotto Malocello, and Majorcan documents speak of "islands newly found in parts of the west," generally interpreted to refer to the Madeiran and Azorean islands. They were likely accidental discoveries from ships caught in the gyre of winds in what has been aptly called the Atlantic Mediterranean. Full discovery had to wait another century, and by then momentum had passed from cities that opened onto the Mediterranean to countries with ports on the Atlantic, beyond Gibraltar. Genoa and Venice and their allies yielded to Portugal and Spain. What began as a rediscovery of Antiquity yielded to the discovery of Earth.[5]

Surely the best-known figure from the full flush of the Renaissance is Leonardo da Vinci. Born in 1452, a year before Constantinople fell to the Ottoman Turks, cutting Europe off from the East and overland expansion, he died in 1519 as the Armada de Molucca under Magellan launched. The Great Voyages of the Renaissance were to geographic discovery what Raphael's Madonnas and *School of Athens*, Michelangelo's sculptures and Sistine Chapel murals, and Leonardo's polymathic notebooks were to literary humanism. In 1452 Iberian mariners had settled most of the restored Canaries and were well along in colonizing previously unknown or redis-covered Madeira and the Azores. By 1519 Vasco da Gama had pioneered a maritime route to India; Afonso de Albuquerque had platted the boundaries of Portugal's empire of the Indies; Pedro Cabral had discovered Brazil; the Atlantic contours of much of the Americas were known; Peter Martyr had coined the phrase *new world* and was a year away from being named official chronicler for Spain's Council of the Indies; the Cape Verde Isles were dis-covered; England and France had launched expeditions to North America; Columbus had overseen four voyages to the Caribbean, two of which made it to the mainland; the pope had divided the world between Portugal and Spain, Hernán Cortés was marching on Aztec Mexico—in brief, the basic contours of the classic Age of Discovery were in place.

The discoveries, and in a few spectacular cases, the conquests, especially the wealth from Aztec and Inca empires, and secondarily the new spice trade through Portugal, unsettled European geopolitics. But discovery had a sim-ilar effect on the state of learning. When he had climbed Mont Ventoux in an echo of a passage from Livy and "to see what so great an elevation had to offer," Petrarch says it happened that his copy of the *Confessions* opened to this passage: "And men go about to wonder at the heights of the mountains, and the mighty waves of the sea, and the wide sweep of rivers, and the circuit of the ocean, and the revolution of the stars, but themselves they consider not." The humanists of the Renaissance turned their talents to just that task of inquiry, what it meant to be human, which is the oldest of human quests, and they looked to the past for guidance and inspiration. Petrarch's ascent of Mont Ventoux is written as allegory. At the top he "stood as one dazed," then he shifted his thoughts "to a consideration of time rather than place," and the journey of mind and soul. "How earnestly should we strive, not to stand on mountain-tops, but to trample beneath us those appetites which spring from earthly impulses." Ovid and Augustine could serve as guides.[6]

The age's explorers took to mountains and mighty waves and encountered worlds unknown to the ancients. They looked to the future, adapting the quest saga to new purposes, and when they climbed Pico de Teide or the peaks of Darien, rather than sink into melancholy over what was lost or re-create a scene described in scrolls written a millennium earlier, they exulted in vast new realms of mountains, waves, rivers, and the circuit of the ocean that extended far beyond the horizons of ancient texts, however newly reborn. Theirs was a complementary task to the humanists, for they were finding whole worlds and peoples unknown to Aristotle and Herodotus. In 1539 the word *discovery* entered English, and in 1543 *exploration*. That same year Copernicus published *De Revolutionibus*, which reconstituted Earthly cosmography. There were unprecedented ways to see land and ocean as well as the stars, and our place within them, and beyond a recovered past a future—the "undiscovered country," as William Shakespeare of the English Renaissance phrased it. The new questers pushed westward past the Pillars of Hercules, the symbolic border to the realm of Antiquity, and never looked back.

Bernardo de Vargas Machuca,
Milicia y Descripción de las Indias (1599),
frontispiece.

BOOK I

Voyages of Discovery

The First Great Age of Discovery

Had there been more of the world, they would have discovered it.
—LUÍS VAZ DE CAMÕES, *THE LUSÍADS* (1572)

By the sword and the compass
More and more and more and more.
—BERNARDO DE VARGAS MACHUCA,
*THE DEFENSE AND DISCOURSE
OF THE WESTERN CONQUESTS* (1599)

The Renaissance Explores

I have inherited from my predecessors a sacred mission. Their labors must not be brought to naught.

—KING MANUEL OF PORTUGAL (1495)[1]

Between 1415 and 1427 Europe did what it had often done in the past. In three incidents it pushed out its borders. None of the sites involved—two islands, one inhabited and one not, and one small peninsula—had been unknown: the contacts were reconnections, and at least two, the inhabited island group and the peninsula, had been known since ancient times. Now, once again, the sails that carried Europe's future billowed amid freshening winds. Yet this time was different. Those sails did not later go slack. Rather, they acquired methods and a momentum that would, amid pauses and pulses, continue for 600 years. Plotting those three points allows for a triangulation over the long arc of European discovery.

The peninsula was Ceuta, on the northern coast of Morocco, a twin to Gibraltar, and so one of the Pillars of Hercules that traditionally defined the limits of the Mediterranean Sea. In 1415 John I of Portugal carried the *Reconquista* across the strait, assaulted the city, took it, and established a beachhead in Africa. One of his sons, Prince Henry (later called the Navigator), distinguished himself in the fight, and three years later beat back a counterattack. What still flickered of Europe's crusading zeal had again crossed the Mediterranean, but this time looking to Africa and the Atlantic and in so doing, in J. H. Parry's terms, also passed "from its mediaeval to its modern phase," a more "general struggle to carry the Christian faith and European commerce and arms round the world." The Portuguese chose to stay—fought to stay. Efforts to push the conquest farther inland repeatedly failed, sometimes catastrophically, but the city's locale made it effectively an island and defensible with coastal fortifications. Eventually Portugal ceded the city to Spain, its Iberian rival, which has held it ever since.[2]

The uninhabited island was Madeira, a two-island group, which may have been mapped as early as 1339 by Genoese-Majorcan mariners, but was effectively rediscovered in 1420 when Prince Henry sponsored an expedition to claim and colonize it. After some initial fumblings, including a wildfire that reportedly burned for seven years and at one point drove the early settlers into the seas, the experiment proved a rousing success. The colony thrived by exporting wood and growing sugar—the onset of Europe's sugar empires. The Madeiran model—a template for discovering, colonizing, financing, and governing—inspired a search for equivalents. In 1427 the Portuguese explorer Diogo Silves correctly mapped the Azores. Madeirans, in turn, helped populate other isles. Between them Madeira and the Azores became Portugal's ports of departure for a global imperium. The islands remain Portuguese.[3]

The inhabited island group was the Canaries. Known to Romans as the Fortunate Isles, lost and found repeatedly throughout the 14th century, they acquired a soft weld to Europe with the return of conquistadors and missionaries in 1402–5, but developed a permanent bond when Portugal took over the project with an armada to Gran Canaria in 1424 and another in 1427. The colonization of the Canaries, however, went slowly and spasmodically. In 1452 Portugal ceded the project to Spain, which required the remainder of the century to reduce the seven islands to its rule.

In one critical factor the Madeira model didn't apply. Madeira and the Azores were uninhabited; the Canaries were populated with an indigenous people who knew the terrain far better than did the newcomers. Not until the demographics changed through war, slaving, and disease did the balance of power turn to the invaders. The Canaries, in their turn, became the principal port of departure for Spanish fleets to the New World; their conquest, the prototype for the conquests that established Spain in the Americas; their economy and government, the model for the much larger projects that became the Spanish empire. They remain Spanish today.

The geography of the islands favored their interlinkages. The Canary current could carry ships down the coast of Morocco to the seven isles, jutting out from Africa like a geologic jetty, but wind and wave made a return by the same means difficult. Instead, ships could swing north and catch westerlies that would take them back to Iberia. Surely, this is how the Madeiras and Azores were discovered in the first place, and it helps explain why Portuguese interest in the Canaries occured just as the Azores were correctly placed. In effect, the winds, waves, and islands made among themselves a loosely contained sea, an Atlantic Mediterranean (much as happened with the North Sea, which might be considered an Atlantic Baltic). Deciphering similar systems over much larger oceans proved essential for blue-water sailing. Early colonies in the Americas projected the triangular trade that characterized the Atlantic Mediterranean across the Atlantic.

The experiences at Ceuta, Madeira, and the Canaries foreshadowed much of the political and historical geography that followed. The Portuguese vanguard. The internecine rivalry between the two kingdoms of Iberia, Spain and Portugal, amid the larger contest between Christendom and Islam. The patterns of contact and settlement. The motives and means of geographic discovery. The pushes and pulls that propelled European outreach in some places and stalled it in others. Nor are the allusions only symbolic: Christopher Columbus, of Genoa, learned his craft amid the triangular trade of the Atlantic Mediterranean. He married a woman from the Madeiras. He departed westward to the Indies from Gran Canaria. He sought new routes, not new worlds.

The Norse had boldly island-hopped across the North Atlantic 300 years before Genoese pilots and Portuguese *marinheiros* circled around the isles of the extra-Mediterranean Atlantic. Then the links broke, the colonies withered, the project became a literature rather than an adventure. Others had visited the Canaries and perhaps additional isles in the past, and like the Norse had retired. This time, however, the Europeans did not. They stayed, and more, they pushed on, or were driven and drawn by others. They launched a Great Age of Discovery whose voyages stitched the waters and lands of the Earth into a colossal patchwork quilt that joined all the world's seas into one.

2

Sails for the Wind to Fill

The How of Exploration

Now you can watch them, risking all
In frail timbers on treacherous seas,
By routes never charted, and only
Emboldened by opposing winds . . .

—JUPITER, OBSERVING VASCO DA GAMA'S VOYAGE TO INDIA,
LUÍS VAZ DE CAMÕES, *THE LUSIADS*[1]

Europe's physical geography had not changed since the great ice sheets had receded. They were borders Europeans had moved beyond before. If they were to leverage the newly found and the refound, a few points of contact in the guise of islands, into ports of departure for an age of discovery, they needed suitable technologies and reasons to apply them. They needed means and motives.

The means included vehicles and the know-how to sail them. The vehicles were fleets capable of both close coasting and far voyaging. The know-how included instruments by which to navigate those ships and weapons with which to defend them, or to use them to project explorers' desires. The two interacted like a self-reinforcing dynamo: better or newer ships demanded better ability to get them where they were supposed to go, and better navigation encouraged further ship development.

Among seafaring civilizations, Europe was anomalous in having two internal seas and two styles of seafaring. A northern tradition centered on the Baltic Sea, with one prong following rivers inland to Russia, the other crossing the north Atlantic, ultimately to the American continent. Ship design favored clinker-built longships and square-sailed cogs, a straight keel, and a centered rudder, all ideal for traffic around the Baltic and its Atlantic extensions. They were ships capable of hauling bulk cargo. The Viking longboat was suitable for raids and the occasional colonizing project, but was better for fighting than transporting commodities like wood and wool typical of the region. To the south was a Mediterranean tradition, itself a hybrid, evolving out of ancient caravel traditions with some loose link to the Indian Ocean. Most recognizable were galleys used both for fighting and trading, but the oar-powered vessel so basic to the seagoing enterprises of Genoa and Venice, while suited to the low-tide, coastal sailing demanded of the Mediterranean and its innumerable internal seas, was vulnerable in the open ocean. Instead, designs borrowed from the Indian Ocean, easily recognized by their slanting lateen sails, from the small dhow to the massive sambuq, were the ones that made it across to the western Mediterranean.[2]

The northern and southern traditions met in Portugal, whose ports opened into the Atlantic and beyond. Here they gradually hybridized into several new species, of which the best known was the caravel. The caravel is a capacious category, more a spectrum of ships than a defined design in the modern sense. Those vessels developed independently of any purposes beyond trade and colonizing along the European and African coast and their adjacent islands. Yet they were, by general acclamation, the best vessels for the great task ahead: the trailblazer, the true explorer, the ship that could most easily move from long voyaging to coastal reconnaissance. But it had little capacity for hauling the supplies a prolonged voyage required, had less sail to speed a ship along, and offered sparse quarters for crews.

No single ship design could do everything. So much as the caravel was a hybrid of northern and southern traditions, so the fleets of discovery were further hybrids with huskier vessels to haul supplies, modifying ships to include more square sails and wider girths, looking part Baltic cog and part Mediterranean caravel. Later, as routes became established, as exploration domesticated into routine trade, ships bulked up into *nau*, carrack, *fluyt*, and later, Indiamen. During the great outburst of exploring and foraging, mixed fleets were the norm. Bartholomeu Dias, who first rounded the Cape of Good Hope, did so with two standard caravels. When he, in turn, oversaw the assembly of the armada that would make the passage to India under Vasco da Gama, he gave it two merchant caravels, both well armed, along with a mixed-rigged caravel and a lateen-rigged caravel. Christopher Columbus had a fleet of three, two caravels (including his favorite, the *Niña*) and a flagship, the *Santa María* (a nau) used for stores. Instructively, it was the *Santa María* that sank off Hispaniola. Ferdinand Magellan, destined for a much longer voyage, had five ships—four carracks and one caravel.

The celebrated vessels of the Great Voyages grew not out of special programs committed to something called exploration and discovery but by tweaking existing technologies to slightly different purposes. That "immensely long passages through unknown oceans" could be made, "and once made could be regularly repeated" by ships "built for the everyday trade of Europe" is, as J. H. Parry aptly put it, "one of the most striking features of the whole story of the Reconnaissance." A century earlier "such achievements would have been unthinkable."[3]

Still, unknown seas were not the only problem. There were unknown peoples as well, some hostile, most wary, all reluctant to cede trade and influence to Europeans without a fight. Its hybrid ships might carry Europeans to new places, but outfitting them with gunpowder allowed them to stay. The classic Mediterranean naval battle was fought with galleys, even outfitted with "castles" fore and aft (the name is revealing) for soldiers to fight from; they were floating battlefields. When gunpowder appeared, it was used like explosive crossbows or spears in the form of arquebuses and equivalent, and cannon were used at sea as they were on land to rake across opposing troops.

The caravel lacked elevated platforms, however, so cannon were mounted along the decks, with gunwales cut into battlements to allow firing, and then—the revolutionary insight—they were installed below the deck. Firing ports replaced the galley's oar holes. Ships became floating batteries that could sink opposing vessels. Eventually, just as store ships (like the carrack) were developed to accompany fleets, so fighting ships (like galleons) were added to defend those fleets. Venice built its first galleon at the Arsenal between 1526 and 1530. Naval warfare veered away from land counterparts, pitting ship against ship rather than armed men against others in shipborne melees. As a goal, sinking ships vied with capturing them. Cannon substituted for large crews, and so proved particularly useful in long voyages.[4]

Naturally, the Portuguese led developments in ordnance as well as ship design, and unsurprisingly, the new firepower was first displayed when Portugal burst into the Indian Ocean. If the caravel let Europeans reach far lands, cannon helped them stay. The great first battles were naval, for as Francisco de Almeida reminded King Manuel, "In so far as you are powerful on the sea, all India will be yours," and without such power it would be impossible to hold lands. The Empire of the East thus concentrated on fortified ports and islands. Eventually, tentatively, Portugal moved inland. But even land warfare was becoming amphibious warfare. Major assaults were launched from ships: these were latter-day Vikings outfitted with armor and gunpowder. They displayed a fierceness that seemed to leap off the pages of chivalric romances. Sometimes a wild surge through the surf was frightening enough to disperse opposing forces, at least unless and until they paused long enough to realize how few the Portuguese were and mounted counterattacks that drove them back to the sea. It was a style and intensity of fighting learned in the brutal border wars of Morocco, even then styled a "school for cutthroats."[5]

That in cameo is the story of Iberian conquests. The shock of surprise and novelty were often enough to prevail in a first fight. Then the Portuguese needed to build coastal fortifications, at which they excelled. The key to victory was to survive the inevitable counterattack, of which there were many. The greatest Portuguese victories were either naval battles (like those at Diu and Hormuz against Turks) or subsequent sieges successfully repelled or, if the fort was lost, retaken. New World Spain quickly erected fortified ports from which the interior entradas set out.[6]

By itself ship design wasn't enough. Almost none of the vessels were built for the blue-water, far-ranging purposes to which they were put. They had to move from shorelines to open seas, from empirically based piloting along coasts to abstract navigation on an unbounded ocean. To operate the hardware the age needed suitable software.

The nurseries of Europe's seafaring traditions were small seas, bounded by coasts or islands, that broke voyages into a series of short sails. Pilots stayed within sight of land as much as possible, or held along fixed lines set by compass or quadrant for spurts across open water. They knew north by compass or by fixing Polaris with an astrolabe (or its simpler variant, the quadrant), though they recognized that the magnetic north deviated from true north and that capturing the north star on a rolling ship could be vexing. Both methods allowed for sailing along a constant line of latitude. (Such methods permitted the Norse to cross the north Atlantic.) Observations could be set against charts or portolan maps, which plotted position along compass lines. Otherwise, pilots relied on an understanding of prevailing winds and currents. Through such means it was possible to traverse east and west along a line of latitude, but not know how far one had traveled, and to know how far one sailed north and south but not along what line of longitude. Navigation was the uncertainty principle of its day. It was "necessarily an art of approximation," and its errors were "cumulative."[7]

The Atlantic posed other challenges. None of Europe's interior seas had much tidal variation, much less tidal streams. The Atlantic had both, and both had to be known in order to sail along coasts and into harbors. The depth of water mattered, too, which made sounding a routine practice and a below-water counterpart to surface charts. Here, again, northern and southern traditions converged. They were ample for continued coasting along Africa, and later, new shores in the Americas.

They were not adequate for traversing the great oceans yet to be found. The pole star vanished at the equator; unreliable compass deviations could be lethal in new seas; nothing of local winds and currents could be recorded until experienced; distances were unknown, or poorly knowable; shorelines could harbor rocks and reefs ready to hole a wooden hull. For the voyages of discovery existing knowledge and navigation devices were suitable to launch. They were not sufficient to ensure the voyagers got to where they wanted to go. They would have to learn, or they would have to borrow from those who already sailed those seas. Part of the success of da Gama's voyage was that he managed to find local pilots, and at Malindi, prior to crossing the

Indian Ocean, probably Ahmad Ibn Majid (who had written the book on Indian Ocean sailing). As the episode illustrates, Europe's achievement was here, as so often, to combine previously disconnected technologies and ways of knowing into a new synthesis. Or as Bailey Diffie and George Winius observe, "The conditions preceding Portuguese expansion into Asia were extraordinary. The only visible assets of the Portuguese were the determination of Manuel, their king, and their new and unique expertise of reaching India by water. The factors opposing them appeared insurmountable."[8]

But the Great Voyages were not simply the result of a technological breakthrough that suddenly made new worlds accessible, any more than the later scientific revolution was the outcome of telescopes and microscopes. Just as someone had to look through those inventions and be prepared to see with new eyes, so someone had to use those tools to new ends, to sail them beyond the known ends of the Earth. The vision lay in the mind. As Parry concludes, "No purely technical explanation, in short, will sufficiently account for European pre-eminence in the discovery of the sea."[9]

Academic lore was of minor value, not merely misleading but often wrong. In any event, the process of geographic discovery was preceded by a process of learned recovery from ancient texts. The Catalan Atlas of 1375 outlined Mediterranean and coastal Europe from portolan charts and Africa based on overland trade, but as with island discoveries, these were lost and had to be discovered anew. In 1410 Pierre d'Ailly assembled much of what had been written about the world into a scholastic summa, *Imago Mundi*, and scholars translated the *Geographia* of Claudius Ptolemy, an atlas that compiled the known world at the height of the Roman Empire. The two books dominated intellectual musings, but neither had much relevance to actual discovery; and Ptolemy's elegant map, plotted along lines of latitude and longitude, "in so far as it was known to seamen at all," Parry notes, "was an almost paralysing discouragement from exploration by sea." Rather, much as the later development of science was a refutation of Aristotle, so "the history of the early discoveries was the story of practical men who proved Ptolemy to be wrong."[10]

The *Imago* and the *Geographia* were not the only authorities. Around 1450, three years before the fall of Constantinople and roughly the time Portuguese explorers reached Cape Verde, Fra Mauro produced an immense map of the world that ranged beyond Ptolemy's (since "more is known") and included among its features a world-encircling Oceanus (the Ocean Sea), not present in ancient texts. These were gorgeous mappae mundi that could inspire learned discourses but had little pertinence for those who had to pilot ships across unrecorded seas to unknown lands.

Knowledge was mixed, as harmful as it was helpful. Much of what was known away from Europe's coasts was worthless or wrong. Maps were full of invisible islands, distorted continents, missing oceans. Mappae mundi offered a stylized geography not much removed from medieval tapestries with their allegorical parables of lions, monkeys, and unicorns. Martin Behaim's celebrated 1492 globe, missing four continents and the planet's largest geographic feature, the Pacific Ocean, had the Earth a sixth its actual size, barely as large as Earth's moon. Elegance substituted for empiricism. There was little to guide a traveler across unknown seas and oceans many times the width of Europe itself.

The same held for traveler tales. Europe had a depository of information from merchants and agents who had traveled to China and India (and missionaries to central Asia and China), over both land and sea. During the lead-in to the Great Voyages, several important new sources appeared. In 1441 Niccolò de' Conti, a Venetian who had traveled widely throughout the east, published an account of what he had seen, identifying the major ports and points of entry throughout the Indian Ocean. In 1487–90 Pedro de Covilhão explored the Red Sea and Persian Gulf, along the east coast of Africa, and across to India. These were reasonably reliable accounts. But for every more or less authentic narrative, there were fantasies like *The Travels of Sir John Mandeville*, published between 1357 and 1371, full of monsters and marvels—exploration's version of the knightly romance. Accounts could be populated with imaginary notions as fully as maps were. There was no authority to sort the false from the true, or the invented from the somewhat askew. There was enough to inspire, if only through avarice and ambition, not enough to guide. Those interested in the Indies, however, understood where they wanted to go. What Europe needed was new ways to get there.

The precipitating crisis was the continued voyaging south by Portugal. Navigators needed a surrogate for the Rule of the Pole Star since at the equator Polaris sank below the horizon. This was a vital concern that could justify a national response. As usual Portugal led, inviting scholars to address the problem, though the notion that a succession of monarchs from Henry the Navigator onward supported a permanent school at Sagres is a myth. Astronomy was too closely allied with astrology, and cosmography, with fanciful rationalizations about how the world must be constituted. Modern science had yet to be conceived; a solution for determining longitude lay another three centuries in the future. But an interaction between practical mariners and scholastics was possible. In combination those who had what today we might call data and those who had skills in manipulating that information into what today might term algorithms could achieve useful results.

Eventually two methods were found. One relied on the readily visible (and inspirational) Southern Cross. With suitable math, it was possible to determine latitude, but the protocols were too complicated for routine use. The other was a more general Rule of the Sun, which used the Sun's altitude at noon to set latitude. The question of longitude was intractable to the methods of the day, and for a while was less critical since, for a time, Portuguese coasting stalled after rounding the great bulge of West Africa. Its marinheiros had found ways to tap into and divert the gold, ivory, and slave trade that had traditionally trekked north across the Sahara, but the coast turned south again and discouraged notions that continued sailing eastward would bring ships to the Indian Ocean and ultimately the Indies.

What happened was a harbinger of an alliance of the practical and the abstruse, of people concerned with making and doing and people concerned with translating and discoursing about texts, that was so characteristic of the Renaissance, here given expression in exploration. In 1484 John II of Portugal assembled a suitable cadre of authorities to translate tables of solar declination into navigational handbooks. One of the group, Jose Vizinho, then voyaged along the Guinea Coast to confirm the results by on-site observation. The experience is an eerie anticipation of what 300 years later occurred with the problem of longitude (in which one of John Harrison's chronometers was field tested by a voyage to Jamaica).[11]

These lessons and learning were consolidated in various ways. One was the amassing of chronicles as Gomez Eanes de Zurara did for Portuguese "Guinea" and Peter Martyr for Spanish America. Cartographic centers emerged in Portugal and Germany. And handbooks were published. The foundational manual (and nautical atlas) of the age, *Regimento do astrolabio e do quadrante*, (Europe's first "manual of navigation and nautical almanac") showed how to use simple instruments to calculate position through latitude—Rule of the North Star, Rule of the Sun, Rule for Raising the Pole (for traversing north and east), along with a theoretical substructure. The oldest printed edition appeared in 1509, ten years after da Gama sailed to India and ten years before Magellan launched around the world. There were schools to train pilots and navigators, and the printing press circulated discoveries, traveler tales, and maps; but research institutions came much later, well after the Great Voyages had sailed into history. Instead, charts, logs, rutters, and chronicles diverged from mappae mundi. They were not synthesized until the late 18th century, when the world's navigable shorelines were at last surveyed.[12]

In short, much as the pioneering voyages were conducted with ships not designed for that purpose, so pioneering navigators had to rely on instruments and maps not invented for their needs. Yet the voyagers made it out and back again and told others how to do it. It should come as no surprise that geographic discovery became, for some leading minds, the model for a more general investigation of the world. (As the voyages expanded, a lot of scholastic learning found itself shipwrecked on the shoals of the previously unknown or misknown.) What mattered was not what could be translated from the ancients but what one could hear from firsthand accounts or see with his own eyes.

3

God, Gold, and Glory

The Why of Exploration

We imagine that we know a matter when we are acquainted with the doer of it and the end for which he did it.

—GOMEZ EANES DE ZURARA, *THE CHRONICLE OF THE DISCOVERY AND CONQUEST OF GUINEA* (1453)[1]

W hy did they do it?

They claimed, in the classic formulation of the age, that they did it for God, gold, and glory. They wanted wealth, fame, and to serve the Cross. Here was the motive power behind the Great Voyages, not the currents ridden by hybrid ships or the revelations from recovered texts or suitable island ports of call. It lay in the hearts and minds of those exploring and those who outfitted them.

But why exploration? Why journeys of discovery rather than trade, conquest, missionizing, and crusading within Europe and its periphery? Why take to the sea, or attack empires many times vaster than the company of adventurers? Or having once voyaged beyond the limits of the world, why repeat the act, and even institutionalize it?

Listing the reasons for Prince Henry's interest in sponsoring voyages of discovery, Gomez Eanes de Zurara notes an interest to know what lay beyond the Canaries and Cape Bojador; the desire for trade and wealth, especially if travels might connect with Christians or lands "into which it would be possible to sail without peril"; to know the extent of Moorish rule "because every wise man is obliged by natural prudence to wish for a knowledge of the power of his enemy"; to find potential Christian allies such as Prester John; and to spread the faith, "for he perceived that no better offering could be made unto the Lord than this; for if God promised to return one hundred good for one, we may justly believe that for such great benefits, that is to say for so many souls as were saved by the efforts of this Lord, he will have so many hundreds of guerdons in the kingdom of God, by which his spirit may be glorified after this life in the celestial realm." All these were traditional ambitions of an ardent crusader, and of particular pertinence to a man like Henry who headed a militant society, the Order of Christ.

But "over and above these five reasons," Zurara noted, there was a sixth "that would seem to be the root from which all the others proceeded, and this is the inclination of the heavenly wheels." Henry's horoscope had reckoned Aries with the house of Mars, Mars with Aquarius (and Saturn), the Sun with Jupiter, and the movement of the heavens with "a certain divine grace" to declare that it was Henry's benign fate to "toil at high and mighty conquests, especially in seeking out things that were hidden from other men and secret." The heavens and heavenly grace bequeathed that Henry should boldly go where no one had gone before.[2]

This is, as biographers have repeatedly noted, the profile of a late medieval prince, still inflamed with the Reconquista, whose whole life would be spent with Andalusia still in Muslim hands and the Moors leering across the Strait of Gibraltar. What distinguished Henry from his contemporaries was his willingness to look to the sea as a means to advance trade and crusading. That task fell to a statesman since "no mariners or merchants would ever dare to attempt it (for it is clear that none of them ever trouble themselves to sail to a place where there is not a sure and certain hope of profit)." Henry himself was as likely to invade overland, usually unhappily, as to invest equivalent energies to seaborne discovery. The voyages were not, in the end, that many or that expansive, and after Henry they stalled, as they had often stalled during his lifetime. Diogo Gomes confirmed that Henry wanted to tap into the stream of gold that otherwise came north across Africa through Moorish traders. In dispatching Gil Eannes to pass Cape Bojador, Henry reportedly proclaimed, "You cannot find a peril so great that the hope of reward will not be greater. . . . Go forth, then, and heed none of their words [prophets of doom], but make your voyage straightway inasmuch as with the grace of God you cannot but gain from this journey honour and profit." What informed one inspired others. "Thus," exclaimed Peter Martyr, "shores unknown will soon become accessible; for one in emulation of another sets forth on labours and mighty perils."[3]

Over and again, others voiced much the same reasons. When asked at Calicut why he had come, Vasco da Gama replied, for "Christians and spices." Christopher Columbus wanted gold ("the Admiral ordered that nothing should be taken, so they [the natives] might know that he sought nothing but gold") and a hereditary fiefdom including a title that he could pass to his heirs, and, like the condottiere of the era, selling their armed companies to the highest bidder, he had shopped his ambitions among all potential sponsors. Looking back on the conquest of Mexico, Bernal Díaz explained that he had joined that corps to "emulate" his ancestors who had served the king, to advance the cause of the Cross ("all our labours are devoted to the service of God"), to win fame (or as Cortés told them, "far more will be said in future history books about our exploits than has ever been said about those of the past"), and of course to become wealthy since "all men alike covet gold, and the more we have the more we want." But ever the practical soldier, Díaz also admitted that there were other, less exalted goads, not least a desperation born of poverty and that once begun a company had little choice but to continue, for as Cortés again reminded them, "it was to save our lives that we had endured all this and worse."[4]

If their reasons seem to echo one another, so they echo those of contemporaries who did not take to sea or carry arms against unknown empires. Explorers did not sail for the sake of disinterested curiosity: they used discoveries to advance purposes widely held throughout the societies of their day and they sought and received comparable honors. God, gold, and glory also animated Renaissance artists, princes, merchants, and humanist scholars. Nor were fame and fortune—status and greed—special to 15th-century Iberia, or the justifications by which the bold rationalize their deeds. Exploration tapped into a reservoir of energies common to the emerging dynamism of a recovering Europe that would lead eventually to the scientific revolution, early-stage capitalism, and overseas empires. Like a small stream whose headwaters cut into and captured a larger one and thus redirected its current, so the inchoate practices that came to be called exploring interacted with the mainstream of late medieval and early modern Europe.

Yet there was, for all this, a ravenous curiosity, a sense that there was something worth knowing around that far cape and over that range of hazy mountains, something not recorded by the ancients, something worth the struggle to find. One could expect wonders as well as wealth. In time a whole new world was unveiled, and an ocean as large as the world envisioned in the mappae mundi of Ptolemy and Mauro, and civilizations as wealthy as Byzantium or Karakorum. Not everywhere of course: there were deserts and mud villages and lands full only of fevers, poison arrows, and penury. There was scant value in untethered curiosity: there had to be a prospect of profit of some kind. But there was a sense of newness and of the unexpected, and a willingness to dare to find what that might mean, a passion for geographic discovery that infected the larger culture, reinforcing trends that would question inherited learning and authority, and that was worth something besides fame and riches. When he saw Tenochtitlán, rising out of its lake, Bernal Díaz thought it resembled something out of a romance and exclaimed that "it was all so wonderful that I do not know how to describe this first glimpse of things never heard of, seen or dreamed of before."[5]

There were pushes and pulls. A centuries old tradition of Iberian border warfare against Muslims suddenly ceased with the final expulsion of the Moors from Granada and left younger sons and soldiers of fortune searching for fresh outlets. The prospect loomed of wealth from colonizing newly dis-covered islands, of intercepting lines of trade across the Sahara by voyaging into the Atlantic and around Africa, of finding and plundering previously unknown empires. Even writing, as Antonio Pigafetta, a Venetian traveling with Magellan, knew could bring glory. "I determined . . . to experience and to go to see some of the said things . . . that it might be told that I made the voyage and saw with my eyes the things hereafter written, and that I might win a famous name with posterity." Great Voyages promised wealth, fame, and adventure, all sanctioned by the propagation of the Cross; and for coun-tries poised at the edge of Europe, the only realistic place to vent such calls came from the sea. Zurara completed his grand *Chronicle* in 1453, the year

Constantinople fell to the Ottoman Turks. Europe would have to go west, not east, and to the west lay the Ocean Sea.[6]

A chivalric tradition and literature found new ways to express itself—new quests for knight errantry, new trials to endure and monsters to slay, new rewards, and in fact fabulous scenes as dazzling as any in romances. When Díaz looked out on the great causeway of Cuitláhuac, across the lakes in the Valley of Mexico, some in his party questioned whether it was "a dream," for as the great towns made of stone rose from the water, it "seemed like an enchanted vision from the tale of Amadís." And who could not see the conquests of Mexico and Peru, the voyages of Vasco da Gama and Ferdinand Magellan, the continent-spanning trek of Cabeza de Vaca as other than epic and romance?[7]

Count nothing impossible! He who will
Always can.
—LUÍS VAZ DE CAMÕES, *THE LUSÍADS*

The Great Voyages were collective projects, the product of institutions, not just the realized wills of individuals. They were state sanctioned, if not state sponsored. They resulted from contests among geopolitical competitors. Nations, too, had their reasons.

What sparked the outburst of discovery in the 15th and 16th centuries was a bitter rivalry between Portugal and newly consolidated Spain. Other states joined in—England, France, the Netherlands, Sweden, and in a different way Russia. But the defining voyages and their character were in many ways an outward projection of a deep dynastic struggle in Iberia. By 1460 Portugal had reached the Cape Verde archipelago when the death of Prince Henry stalled further expeditions. By then Iberia was divided into three kingdoms. Aragon looked to the Mediterranean; Castile held the interior, save for Moorish Andalucía; and Portugal looked to the Atlantic. The three states quarreled for supremacy in Iberia, and Castile and Portugal skirmished down the African coast, with the Canaries a special prize.

In 1469 Isabella declared herself the heir to the Castile throne. Had she married the widower Alfonso V of Portugal the history of exploration would have gone one way; if Ferdinand of Aragon, another. She married Ferdinand, which set off a war of succession in which at one point Portugal ineptly invaded its rival even as it renewed exploration around the bulge of Africa. Not until 1479 did the war conclude with Castile and Aragon joined, and the Treaty of Alcáçovas calming the conflict between the new state, Spain, and Portugal. In Iberia Spain and Portugal remained separate. Spain promptly laid siege to Andalucía, completing the Reconquista in 1492. In the Atlantic the two crowns divided the discovered islands between them. Spain got the Canaries, which it sought to reduce by repeated invasions, not subduing Tenerife and La Palma until 1495. Portugal retained Madeira, the Azores, and the Cape Verde Islands, along with rights to new lands and islands beyond Guinea where it laid down stone *padrãoes* as claims. In 1482 it established the fabled fort at São Jorge da Mina halfway across the Bight of Benin. Each state's islands became the ports of call for an alternative silk road of maritime trade and discovery. The division of the Atlantic Mediterranean prefigured the division of the lands and loot from the voyages that followed. Between them Spain and Portugal inscribed the political, commercial, military, and cultural matrix for the First Great Age of Discovery.

Competition continued. Spain and Portugal quarreled over the coming centuries for supremacy, eventually just autonomy, and of course for prestige, in Iberia. But overseas trading factories and new discoveries allowed both another arena for rivalry and an outlet. Iberia did not have to simmer in its own juices: it could export its soldiers, merchants, priests, and ambitions. If one party faltered, the other would step in. If one crown did not support an explorer's scheme, the other might. Christopher Columbus proposed his westward voyage to the Indies to Portugal, then the leader in maritime exploration, and when it demurred, he took it to Spain. Ferdinand Magellan, after Portugal rejected his expanded scheme for westward voyaging, took it also to Spain. In 1494 the Treaty of Tordesillas divided the globe as the Treaty of Alcáçovas had the eastern Atlantic. The competition extended to the Philippines and the Spice Islands.

In Europe's previous expansions, what had pushed out was pushed back. This one kept going, and one critical cause was the pressure of rivalries. It was not enough to compete against an external rival—animating as that might prove. What mattered was internal competition within and among European states. The fact is, *Europe* did not discover the rest of the world; particular European states did. Europe had little political unity; Christendom had meaning only against the Islamic caliphate; European states were constantly squabbling among themselves. After the Reformation, even Europe's religions sparred for influence and carried their rivalries throughout the world. Yet that ceaseless turmoil kept the pot boiling. Internal fights were projected outward. If Portugal stumbled, Spain strode into its place. And when the Iberian dynamic finally stalled, other countries were ready to poach their overseas possessions. What could not be accomplished internally in Europe might be achieved overseas. As long as Europe was politically unsettled, partitioned among quarrelsome fiefdoms—and it remained preternaturally ill at ease—there was support for exploration and all that accompanied it.

To the question posed by so many chroniclers of the enterprise, Why did they do it?, one answer is that, if they didn't, someone else would. Probably someone they didn't like and didn't want to succeed. That, too, became a reason to act.

Even so, for every swaggering conquistador, there was a critic. Among explorers, there were hardships, some unspeakable, only a fraction of which were rewarded. Again, Bernal Díaz spoke for thousands when he said simply, "Such as the hardships to be endured when discovering new lands in the manner that we set about it! No one can imagine their severity who has not himself endured them." But enduring suffering for a noble cause was different from questioning the cause itself.[8]

Perhaps the most astonishing critique came from Luís Vaz de Camões, the poet laureate of the Portuguese empire who had spent years in the East. In *The Lusíads* he attributes the wind in the age's sails coming from the "giant goddess Fame" and from the gold that "conquers the strongest citadels," that turns friends into "traitors and liars," debauches nobles and maidens, and that can buy "even scholarship." The great navigators will rival the exploits of Odysseus and Aeneas. But they will also cause grief and pain. In one of the most moving passages of the epic, he imagines da Gama's fleet as it unmoors from the piers at Belém. The crews try to avert their eyes from the sight of loved ones left behind and to close their ears to those petitioning them to stay. But they cannot avoid an old man, his eyes "disapproving," who harangues the departing ships from the shore, hurling prophecies and mockery from a wisdom plucked out of a "much-tried heart."[9]

One by one the Old Man of Belém demolishes the vaporous presumptions that fill the sails. Honor is no more than "popular cant"; fame, but vainglory; "visions of kingdoms and gold-mines," delusions; bold discoveries, mere folly; idealism, a disguise for greed. Crusading zeal is better satisfied closer at hand. Adventuring will only lead to "new catastrophes" and wreck "all peace of soul and body." The glitter of gold is a seductress's call that will deplete rather than enrich. Subtly, yet "manifestly," the *carreira da Índia* will "consume the wealth of kingdoms and empires!" The voyage is but the latest example of an interminable, tragic restlessness.

> In what great or infamous undertaking,
> Through fire, sword, water, heat, or cold,
> Was Man's ambition not the driving feature?
> Wretched circumstances! Outlandish creature!

There lies the meaning of the Old Man's lament: the unalterably flawed character of humanity, because of which the founding epic must match triumph with tragedy.[10]

The chivalric impulse so ardent among the adventures could turn sour, or in the hands of Miguel de Cervantes Saavedra, to satire. Cervantes had proudly fought in the wars against the Turks (losing the use of his left arm at the Battle of Lepanto), and endured slavery in Algiers, before working for the state as a tax collector and suffering imprisonment for suspicious accounts. He was, thus, a Spanish counterpart to Camões, and while like him proud of his trials and service, he also saw the dark side of empire. In two volumes, one published in 1605, the other in 1615, he tells the tale of *Don Quixote*, a man inspired by chivalric ideals, but also deranged by them. It's a gentle satire, but a critique nonetheless. If Quixote had gone to New Spain, the tale might easily have turned to madness.

It might, perhaps, have resembled the picaresque tale of Fernão Mendes Pinto, a Portuguese whose *Perigrinaçao* (*Travels*), an alloy of adventure, exaggeration (including some probable fabrication), and allegory, at once celebratory and cautionary, recounts his adventures in the Enterprise of the Indies. "It seems," he wrote, "that misfortune had signaled me out above all others for no purpose but to hound me and abuse me, as though it was something to be proud of." Yet he confessed how "God always watched over me and brought me safely through all those hazards and hardships," such that "there is reason to give thanks to the Lord," for there are no misfortunes so great that human nature, with God's help, cannot overcome." If *The Lusíads* is a kind of Portuguese *Iliad*, Pinto's *Travels* are its *Odyssey*.[11]

Born into rural poverty, Pinto arrived in Lisbon with an uncle in 1521 (the year Magellan's remaining ship returned). The coming years were unhappy, and in 1537, aged 28, Pinto decided to go to India. As he ruefully noted later, it was not fortune that carried him to the Indies; it was a search for fortune that sent him there. It was easy to find passage; the Estado da Índia was chronically short of Portuguese. Pinto had one goal: "to be rich, which is all that I cared about at the time."[12]

The passage to India put Pinto at Mozambique; Massawa; Ethiopia; Diu; Hormuz; Goa; Mallaca; Lugor in the Gulf of Siam; Burma; Java; Macau and Canton; Okinawa, Japan, where he made four journeys; the Red Sea; the Persian Gulf; the Indian Ocean; the Arabian Sea; the Bay of Bengal; the South China Sea; and the East China Sea. He faced battles, storms, shipwrecks, piracy, and slavery, and underwent a religious conversion. He traded. He became wealthy. He joined the Jesuits, donated his fortune, then withdrew. Finally, after 21 years, he returned to Portugal in 1558, expecting to be honored and rewarded for his services. Nothing happened. He wrote his *Travels*—for his descendants, he claimed. It was published 30 years after his death in 1614—the year Cervantes completed *Don Quixote*. By then the Estado da Índia had already begun its painful, inexorable implosion.

It's an astonishing, picaresque tale, and even if much is exaggerated and some adventures likely invented, they show the reach of Portugal's eastern empire, and reveal vividly, satirically, the real motives behind it. Like Pinto, Portugal had voyaged East mostly to become rich. There is little surprise that it could not hold what it had intended to loot. Asian and African powers reclaimed some sites, and the trade overall, but the deeper wounds came from European rivals. Year after year, they picked off one base after another such that in time the passage to India became a near Dutch monopoly, until the British did to the Dutch what the Dutch had done to the Portuguese. But the worst damages were self-inflicted. In the end Pinto blamed his misfortunes on his own sins. Like Pinto too many of the Portuguese came to plunder, and the national wealth accrued was soon squandered, in the classic way of ambitious monarchs, in pointless foreign wars. Like Pinto, Portugal was left in financial and then political penury. The callow King Sebastian plunged into Morocco, where both he and his army were slaughtered. He left no heir, and in 1580 Portugal was joined to Castile under Philip II. Philip then squandered his good fortune with foreign war quagmires in the Netherlands and an ill-fated armada to England.

The Portuguese empire lingered on for another 400 years. But Portugal might wonder what, apart from fame, it had gained.

Losses could and frequently did overwhelm rewards: voyages could experience an appalling attrition of life and ships. For every returning hero there were scores who did not return at all. There were, after all, other outlets for wanderlust, fighting, and profit seeking. In his *Chronicle of Guinea* Gomez Zurara records the experience of Dinis Eannes, who exhorted his band that "all the amount of our gain dependeth on our labour," and that all had agreed to go. They found nothing. "So they returned again, not without great weariness; for what they felt most sorely, after going through such great toil, was the finding of nothing that they had sought." Yet somehow, the stars and planets aligned in ways that inspired a generation to keep trying anyway.[13]

Surveying the European discovery of the Americas some 540 years after Gil Eannes tacked past Cape Bojador, Samuel Eliot Morison, dean of early American exploration history and a biographer of Columbus, posed the core question. "Why did they do it? Why did they risk it? Why did they keep going back?" Morison could offer no definitive answer. "I wish I knew."[14]

4

Where No Human Being Ever Sailed

The What and Where of Exploration

"What brought you to this other world
So far from your native Portugal?"
"Exploring," he replied, "the vast ocean
Where no human being ever sailed."

— LUÍS VAZ DE CAMÕES, *THE LUSÍADS*

The era of the Great Voyages was broad in its geography and brief in its history. In less than a century from the death of Prince Henry, or half a century from the time Bartolomeu Dias rounded southern Africa, emissaries from Europe visited five of the world's continents (possibly six) and six of the world's oceans. They found that the Earth's seas were one. They traced the contours of Earth's continents sufficiently to navigate around or through them. They linked the major maritime civilizations. Lisbon connected with Nagasaki, Seville with Tenochtitlán, the Strait of Gibraltar with the Strait of Malacca. When the old wisdom broke down, everything seemed possible. Ships blew out of Europe like shrapnel. It was an astonishing accomplishment for what might rightly be considered a single generation.

Discovery and navigation assumed two forms. One took the model of the Mediterranean and Baltic—small seas joined by straits—and sought equivalents first in the Atlantic, and then across the globe. This meant the seas could be joined; that, ultimately, a ship could put to sea at one port and end up at any other. The other process deciphered the patterns of wind and currents. This meant ships could move from piloting to navigating, that they could let coasts recede from sight, take to the unmarked sea, and turn routes out into routes back. Wooden sailing ships could ride prevailing winds and waves in loops, not simply hug the littoral. Great voyages, not simply long ones, became possible.

PORTUGAL: *CARREIRA DA ÍNDIA*

The era began with the rediscovery of Atlantic isles and coastal trading along the western coast of Africa. Portugal led, often jostling with Spanish interlopers. That it might be possible to circumnavigate Africa seemed likely when the western bulge bent eastward. But the goal was to intercept the Sahara trade in gold and slaves closer to its source. The great trading fortress at São Jorge da Mina (Elmina) did that. The prospect for continuing around Africa stalled when the shoreline bent southward, and continued for hundreds of miles of tricky coasting. In 1488 Bartolomeu Dias, blown southward and eastward by days of storms, found himself on the other side of Africa and in the Indian Ocean. The long-sought passage to India was at least possible. Still, it took another decade before Portugal could muster the materials and political will to complete the voyage under Vasco da Gama. The old dream—the dream of crusaders, the dream of traders—to circumvent Islam had happened.

It was a difficult passage. The African coast was hostile and dangerous, winds and currents were at odds with intentions. Success came by understanding that the gyre of currents and winds in the south Atlantic was a reverse image of those in the north. When da Gama sailed, he turned away from the coast to the open sea, caught the trade winds to carry him southwest, and then turned and filled his sails with the trades to blow him southeast in what became known to the Portuguese as the *gran volta*. Decoding the winds was a reasonable guess, and an inspired one. Had Portuguese caravels continued to crawl along the coast, it's unlikely they would have turned that hazard-plagued voyage into a regular trade route.

Nor if they had not understood what awaited them. They did not have to sail unknown seas to India. Rather, they had to creep along the east African coast until, between Kilwa and Malindi, they encountered the thriving maritime traffic that rode the monsoon winds between Africa and India. Here they did not have to discover or invent sailing routes: they were embedded within them. They found local pilots. What followed was a process of translation, not easy, not without errors, even fatal mistakes, but a different project than turning a ship to ride the wind and see where it might go.

Once in that realm they could join the existing flow of seaborne trade. They could locate the vital straits, pass through to new seas, visit known and new islands. Doubling the Cape of Good Hope brought the Portuguese to the Indian Ocean and its well-oiled trade network. The Strait of Malacca took them to the South China Sea, the Spice Islands, and then to China and Japan. Some of this they previously knew, or knew as rumors and speculation; after all, such putative knowledge was what drew them to explore new routes. Once on the scene they learned from existing travelers where to go and how to get there. They knew, or quickly learned, those straits that favored competitors such as the Red Sea and the Sea of Hormuz and sought to block them.

What humanist scholars were doing in Europe, translating ancient texts, explorers did for geographic lore. They used native guides, indigenous pilots, and formal texts for navigating the seas of the Indies, and recoded them for European civilization. With few exceptions they did not discover facts unknown to humanity; they incorporated the scattered lore of maritime peoples into a collective cartography. They could move fast because they had only to reach the next sea and then tap into local lore. Da Gama's first landing party at Calicut found an Arab who spoke Castilian. Columbus seized six natives at his first landing, in San Salvador, and released them to communicate with locals at each subsequent island; offshore at Honduras he found a large trading vessel from the north. In short order, the world ocean resembled a colossal patchwork quilt stitched together by Europe's mariners.

The chronicle of Portuguese voyages startles in its pace and boldness. Marinheiros mapped the northwest bulge of Africa by the 1470s, including the coastal isles São Tomé, Príncipe, and Fernando Pó. By 1488 they had completed a reconnaissance of the entire west coast. They were in the Americas by 1500—Gaspar Corte Real in Newfoundland (though Portuguese fishermen surely preceded him), Pedro Cabral to Brazil. Along the carreira da Índia they sighted other Atlantic isles—St. Helena (1502), Ascension

(1501–3), Tristan da Cunha (1506). They surveyed the eastern coast of Africa. They arrived at India in 1498, and sighted assorted islands in the Indian Ocean—Madagascar (1500), the Mascarenes (1507–11), Diego Garcia (1512), the Maldives and Sri Lanka (1518). In 1511 they had reached Malacca and seized the strait. They reached the fabled Spice Islands a year later. They were at Canton and Macau by 1513. They contacted Japan in 1543. Once in the Spice Islands and the East Indies proper they would surely have learned of the seasonal trade between them and northern Australia. It's possible they made it to Australia and even speculation that, later chasing Magellan, they circumnavigated the continent. Fifty years after Columbus announced he had reached Japan, or at least its outer islands, the Portuguese actually had. No less astonishing, they returned, and sent subsequent vessels to retrace the routes until the Great Voyages of discovery morphed into routine traffic for trade. Luís Vaz de Camões was right. Had there been more of the world they would have discovered it.[1]

This was the master story at the time: Portugal had pioneered a sea route to the Indies, even to the Spice Islands themselves. They had encountered two new continents (the Americas), and possibly a third (Australia), confirmed the dimensions of Africa, and begun the task of revising ancient cartography to accord with modern data. For a principality on the fringe of a continent itself on the fringe of Eurasia, it was an astounding achievement. It established the foundations for an empire of trade and settlement, portions of which survived under Portuguese rule until the end of the 20th century. Some discoveries such as Madeira and the Azores have remained a permanent part of Portugal; the Portuguese empire of the Indies survived as long as the western Roman Empire had from Augustus to Romulus Augustus.

When, from time to time, someone argued for a western route to the Indies, officials met their claims with suspicion and scorn. Portugal had a working route, it had no need for another, and did not welcome the thought of competitors. Equally, it had learned something of the actual distances involved and did not accept, as advocates like Columbus did, Marco Polo's exaggerated length for Asia, Claudius Ptolemy's diminished dimensions for the planet, or Martin Behaim's blinkered globe. They were confident Columbus had not brushed against the Spice Isles but rather had found yet another batch of Atlantic isles. And for his first voyage they were right.

SPAIN: *ORBIS NOVUS*

But if correct about Columbus' murky geography, Portugal was wrong about the significance of the western seas he sailed and the lands they held. The West Indies and the mainland behind them offered a counterweight to the East Indies for Portugal's blood rival, Spain, which found greater wealth in conquering the Aztec and Inca empires than Portugal got from Goa and Malacca. Portugal found a new way to connect the Old World; Spain found a way to connect the Old World to a New. The carreira da Índia changed the balance of power in Europe. The voyages to what became known as the New World changed the balance of world history.

The project began with Christopher Columbus, a Genoese pilot, trained in the Atlantic Mediterranean circuit, who proposed to reach the Indies by sailing west. A prickly personality, eyes glazed with gold, he was an able pilot, a poor geographer, and a miserable administrator of the lands he found. There is little evidence that he accepted the magnitude of his real discoveries, believing with little proof that he had in truth found a westward path to the east and that the islands he explored lay off the coast of Japan. Others understood his achievement better than he, yet over the course of four voyages he redirected Spain's energies from the East Indies to what came to be called the West Indies, which had a maritime culture that quickened his reconnaissance of the entire chain.

He departed from the Canaries and returned via the Azores: in effect, he added another sea (Oceanus) to the maritime mosaic by which European mariners had pushed beyond the continent's shores. His first voyage (1492–93) brought him via the trade winds swiftly to the Bahamas, and then to Cuba and Hispaniola. They became the Antilles, after one of the Atlantic's mythical islands. His second voyage (1493–96) explored more of the Caribbean chain, in this case the Lesser Antilles. His third voyage (1498) took him to Trinidad and the mouth of the Orinoco River, sighting the mainland of what appeared to be a large landmass (*"otro mundo"*), not previously known. His fourth voyage (1502–4) allowed him to complete a crossing of the Caribbean and a coastal reconnaissance of much of Mesoamerica. During his second voyage, Bartolomé Colòn crossed Hispaniola, foreshadowing the more monumental crossings of lands to come.

No comparable explorer of this era discovered so much through so many distinctive voyages.

His career ended less gloriously than it began. After his first voyage, it was said that everyone wanted to go to the new islands. After his second, no one did. He returned to Spain in chains for misgoverning the settlements on Hispaniola. His fourth voyage concluded with him and his crew marooned on Jamaica. Lured by gold, the promise of pearls, and the mirage of the Indies, it was left to others to appreciate far better than he what he had discovered and turn it to profit. He swiftly translated the indigenous knowledge of the Antilles, though into what he wanted to hear. All in all, Columbus not only sketched the basic geography of the Caribbean Sea, but established the character of the Spanish presence and founded the ports from which the conquest of the mainland would be launched. The history of Hispaniola became a sad cameo of what followed.

Others filled in the blank spots, dazzled (as Columbus had been) by the prospect of gold and glory, but now able to depart from Hispaniola not Seville. The Gulf of Honduras, the Yucatan Peninsula, the Gulf of Campeche—all were mapped. Two expeditions for which Amerigo Vespucci was a member or leader (1497–1500) extended the mainland to the mouth of the Amazon River; another two expeditions for Portugal further explored the coastline of Brazil (which Portugal claimed), and got Amerigo's name attached to the New World. In 1508 Cuba was circumnavigated. In 1512 Vasco Núñez de Balboa trekked across the llanos and northern cordillera; a year later he crossed the Isthmus of Panama and spied the Pacific Ocean (which since he was looking south he named the South Sea). That same year Ponce de Leon traced the contours of southern Florida. In 1519 Hernán Cortés began the conquest of Mexico, and Ferdinand Magellan completed the coastal map of South America's southern cone, including his eponymous strait. The conquest of Aztec Mexico set up a breakout of overland expeditions into North America and Mesoamerica, inspired by a quest for similar "Mexicos" elsewhere in the New World. After several coastal surveys, Francisco Pizarro found one in Peru in 1532. That set off another explosion of exploration by conquest throughout South America.

The major dimensions of what became New Spain were fleshed out in the 1540s. In South America Francisco de Orellana crossed over the Andes and sailed down the Amazon River to the Atlantic in 1541–42, while Pedro de Valdivia passed over the southern Andes to the outskirts of present-day Buenos Aires. At the same time in North America Francisco de Coronado trekked overland from Mexico through the Southwest and Hernando de Soto through the Southeast. Between them they discovered the Colorado and Mississippi Rivers, the Grand Canyon, and the Great Plains, but no new El Dorados. By the time Portugal had mapped the Greater Indies and reached Japan, Spain had discovered two seas, one ocean, and two continents.

With Spain, as with Portugal, it had happened over the course of a lifetime. When Hernán Cortés was born in 1485, Spain was jousting with Portugal down the coast of West Africa and trying to reduce the last of the Canary Islands. When he died in 1547, Spain had explored and brought under imperial claims the Caribbean Sea and its islands, and two Amerindian empires, the Aztec and the Incan. It had explored the southern half of North America and traced the coastline of South America. A band of survivors under Cabeza de Vaca had trekked from Florida to Mexico. Eager conquistadores had crossed South America at both its widest and narrowest. In the Age of the Reconnaissance, in almost every respect excepting only the far north, Iberians led.

THE REST OF THE RECONNAISSANCE

They did not leave a lot for anyone else, they were keen to keep others away, and they did not release critical information. Discoveries—costly, arduous—were trade secrets, if not state secrets. Spain housed them in the Casa de Contratación de las Indias, the House of Trade; Portugal, in the Teosoro, the Treasury. What remained was to probe some unattractive routes, mop up some inhospitable corners, try to outflank the protected trade routes of the first movers, and to poach on Iberian claims. A few nations did both, and one, Russia, pioneered a different style of geographic discovery necessarily based on land.

North America drew the most attention. It's possible, probably likely, that the great cod fisheries of the Grand Banks were known and visited by fishermen before formal discovery. If not, they were soon recognized and attracted Portuguese, English, and French fleets. Formal exploration began when John Cabot, another freelance Italian pilot, moved to Bristol, England, about the time Columbus returned from his first voyage. Like Columbus he believed a westward passage to Asia was possible, and further asserted that a northern route would be shorter. In 1496 Henry VII granted him rights similar to those of Columbus; the next year he sailed the *Mathew* to the North American coast, probably Newfoundland, which, again like Columbus, he insisted was Asia. He returned the next year with a fleet of five vessels, one of which limped into Ireland while the rest, with all hands, including Cabot, disappeared.

The task fell to Portuguese freebooters from the Azores. In 1499 João Fernandes sailed to the mainland, and between 1500 and 1502 three brothers, the Corte Reals, managed a series of voyages that traced the coast from Labrador to Nova Scotia. Meanwhile, more Azoreans, beginning with Fernandes, offered their services to Bristol interests. With Cabot's son, Sebastian, they sailed annually from 1501 to 1505 and affirmed what the Corte Reals had found, that this was not a northern Antilles, but a mainland. The unwelcome news was that it was not Cathay, not a northern land of spices, not part of a maritime trading circuit. What the sea had was cod—a fabulous fishery. What the land had was fur and timber. By now Portugal was obsessed with its Indies trade and withdrew, save for its fishermen. Some seasonal settlements sprang up for the cod trade, populated by English, French, and Portuguese.

Meanwhile, the Armada de Molucca, launched under Ferdinand Magellan, completed by Juan Sebastián Elcano, returned with the revelation that a strait existed through South America. That inspired the northern aspirants to search for a comparable passage through North America. In 1524 France dispatched another Italian, Giovanni da Verrazano, to the northern New World, where he sailed from Newfoundland to Florida; and Spain pursued the identical strategy, but from the south, this under a Portuguese pilot, Esteban Gómez, who covered the same shores but in the opposite direction. They found no strait, only a long, imposing mainland. If there was a passage, it would be found in the icy sea far to the north. A decade later Jacques Cartier tried, and found the watery entry to Canada, the Gulf of St. Lawrence and the St. Lawrence River. As a passage it led to the interior, but not through it. Over the coming decades some less ambitious voyages were undertaken, a few leaving settlements. Like the Norse before them, none survived. North

America was not a passage to India, nor a Mexico or Peru, not even a Hispaniola. Not until the onset of the 16th century would efforts to locate a strait through the ice renew, and not until the onset of the 17th century would efforts to establish trading posts or settlements revive.

But if a northwest passage stalled against a wall of ice, and a Little Ice Age, perhaps it would be possible to try the northeast. If a northwest passage was the flip side to the Strait of Magellan, a northeast passage was an inverted Cape of Good Hope. Perhaps there was a northern carreira da Índia, one shorter than the interminable voyage around Africa. If the North Sea was an Atlantic Baltic, and the circuit between the Canaries and Madeira an Atlantic Mediterranean, perhaps the White Sea would be a polar Baltic, part of a mosaic of new seas and straits that could carry stout traders to the riches of the east.

Russians and Norse had been in the waters, but their experiences were not incorporated into formal knowledge. Someone else would have to absorb and codify the northeast's harsh geography, and they would likely undertake it somewhat out of step with the Great Voyages. So it was during the 1550s, when Portugal and Spain were consolidating their empires, that Holland and England began exploring a northern route around Eurasia. Regional attractions existed, notably the fur trade centered in Muscovy. In 1553 England chartered a Muscovy Company to support trade and furnish a vehicle for discovery; and for the next 30 years it sent voyagers around the Kola Peninsula and into the White Sea, and possibly touching on the Kara Sea, from where rivers and portages could take it to its goal. The sea opened only briefly, however, at the end of summer, which left most expeditions to winter over. They were brutal months, from which many did not survive.

The Dutch then took up the quadrant, with a series of voyages under Willem Barents. Barents succeeded in tracing the northern littoral of Novaya Zemlya and left his remains on its harsh coast while wintering over, but also deposited his name on the sea between it and the Kola Peninsula. Commercial links were established between Europe and Muscovy, but there was no northern route to the Orient. In 1619, suspicious about interlopers, the tsar closed the eastern approaches through the Kara Sea and Ob River. He thus shut down the prospects for connecting Europe's maritime voyages of discovery with the overland expansion eastward driven by a vigorous frontier of *promyshlenniki*, effectively Cossacks of the fur trade.

The Mongol invasions had hammered medieval Russia, which found itself blocked on the east by the steppe hordes and on the west by a reinvigorating Europe. Over centuries the hordes had weakened, then melted away; the passage east became far easier than expansion westward. At breakneck speed Russians sprinted across Eurasia by exploiting its rivers. The land was vast, snowy in the winter and boggy in the spring; rivers offered boat passage in the summer and ice corridors in the winter. The promyshlenniki quickly found how to link one river system to another—a land-based version of what was happening on the world ocean. In 1581–82 Yermak Timofeyevich crossed the Urals into the Land of Sibir. By 1632, while Boston was being platted, Cossacks had reached Yakutsk; by 1639 Semen Dezhnev rounded the Chukchi Peninsula and coasted along the Kamchatka Peninsula and crossed by the Sea of Okhotsk; by 1651, expeditions under Vassili Poyarkov and Yerofey Khabarov had sailed down the Amur River to the Sea of Okhotsk; by 1679, they were on that sea's coast at Udek. By then much of the interior river system was explored, filling in such prominent features as Lake Baikal (1652) and the coast of the Sea of Okhotsk. It took the wishes of the dying Peter the Great to send Vitus Bering to the strait that bears his name—the last significant passage between the world's oceans to be discovered. The old dream to find a western passage to India had been reversed, as Europe found an eastward passage to the Americas.

Here as throughout the European imperium the newcomers relied on natives for the production of wealth. A fur tax prodded the quest for new lands. The rapidity was breathtaking. In some 57 years Cossacks and fur trappers had bolted across Asia, and traced the contours of what for a northeast passage might be the Cape of Good Hope; in fact, the Arctic passage was a frozen desert, far worse than Dias's Cape of Storms. Yet the velocity of discovery on rafts and dugouts and sleds is comparable to that for the Iberians in their caravels. Not much became known to the rest of Europe, however, for Russia was ever reluctant to advertise to potential adversaries its discoveries and imposed closure to outsiders, as Spain did in the New World. Throughout, Russia seemed a half step ahead or behind the rest of the Age of the Reconnaissance.

If the 16th century was the era of bold long voyages and rude geographic outlines, the 17th century was an era of filling in blank spots and of new rivals nipping at the heels of the Iberians or testing the lands just outside the Iberians' reach; the infilling was the geographic counterpart to dictionaries and botanical systematics. Piracy and poaching helped spread the new discoveries beyond their nominal rulers. The English renewed probes for a northwest passage; the French, the St. Lawrence River and environs; the Dutch, in North America's northeast and Brazil (which it seized from Portugal from 1630 to 1654); and throughout the passage and empire of the Indies. A handful of captains repeated a global circumnavigation; excepting Francis Drake's voyage (1577–80) that managed to intercept the annual treasure ship from Mexico to the Philippines and, thanks to a storm, found the passage south of Cape Horn, most proved discouraging. There was an immense sea, a dangerous route around South America, and little profit to compensate for hazard and suffering.

The English and Dutch picked off Atlantic islands like St. Helena and Ascension; planted rival trading posts along the carreira da Índia; and took on the Portuguese with trading factories in India, Japan, and the Moluccas. Keen to avoid direct conflict where possible, the Dutch experimented with new routes in the Indian Ocean, a kind of *gran volta*, that took them away from the Portuguese route along the coast and to the eastward, where they would catch trade winds to take them north to India, Batavia, and coastal Asia. Sometimes storms pushed them farther than they wished, and they found new islands and even a continent. From 1618 to 1627 Dutch voyagers traced the western coast of Australia. Between 1642 and 1644 Abel Janszoon Tasman virtually circumnavigated the continent, sighting Tasmania before discovering New Zealand, and then swinging north through Melanesia and Indonesia, and later surveying Australia's northern coast.

The new rivals erected trading posts and small settlements in North America, outside the reach of Spain. The English and French created a long rivalry, powered by fish and furs. With Hudson's Bay Company, the English platted a series of factories around the bay that formed a kind of fur watershed. The French emulated the Russians and exploited rivers and portages to dash around the English in a long loop across the Canadian Shield. The main job of discovery, however, had to wait for the next century when their North American colonies were entangled with a new Hundred Years' War between France and Britain.

All this was a process of filling in what was known often in outline. By the mid-17th century the Great Reconnaissance was complete. Europe certainly didn't find all there was, but it did find all it wanted. Countries had turned to exploration to boost trade, power, and prestige, and as a surrogate for Europe's internal geopolitical quarrels. The goal had been trade routes to the Indies; and after Spain's conquests in the Americas, a parallel search for El Dorados and Mexicos. Yet by the end of the 17th century, it had accomplished that task. The routes to bind Europe to the rest of the world were known. There were no new worlds to discover. Exploration had morphed into trade, tourism, and piracy. It was easier to replace another's trading post than explore for a new source of trade. Europe had reestablished its preferred geopolitical order, based on a balancing of powers. There was little reason to continue. Discoveries occurred at the margins.

When it launched, the Great Reconnaissance had brisk winds and a following sea. Two centuries later it either sailed with the trades or found itself becalmed.

5

Isles

The Brave New Worlds of Discovery

Miranda: "Oh, wonder! How many goodly creatures are there here!
How beauteous mankind is! O brave new world
That has such people in't!"
Prospero: "Tis new to thee."

—WILLIAM SHAKESPEARE, *THE TEMPEST* (1611)

In retrospect, the great discoveries seem to be about old and new worlds, about unknown oceans and unimagined continents. In the 15th century they were mostly about islands. For seaborne discovery continents were barriers; islands, protected ports of call.

Islands were sanctuaries in an unpredictable sea, sites for provisioning and repairs, points of departure for further exploration, fortifications for defense amid often hostile rivals, new sites for colonization, and prods for the imagination. They were the beginning of a voyage and its end and the rest stations along the way. Columbus left from the Canaries, sailed to the Antilles, and returned via the Azores. Da Gama departed from Madeira and returned by way of the Cape Verdes and Azores.[1]

Points of departure—Madeira, the Canaries—had their counterpart in off-shore islands from São Tomé to Hispaniola or in seaports that had the properties of near islands such as Elmina, Goa, Zanzibar, and Mombasa. If a new land did not have a port, the newcomers promptly built one like those at Vera Cruz and Lima. Hernán Cortés subdued Tenochtitlán, islands in a lake, by building boats and attacking it as by sea. The Indies, East and West, are a concourse of islands. Even in the 17th century the Spanish monarchy proclaimed the Canaries, located at a triple junction between Europe, Africa, and the Americas, as "the most important of my possessions, for they are the straight way and approach to the Indies." In the charter (*Capitulations*) Columbus obtained, he asked to be made viceroy and governor-general of any islands he should "discover or gain." Sancho Panza mockingly begged Don Quixote to reward him with the governorship of an island.[2]

On maps of the time islands loom large: charts were maps of islands in the Mediterranean, islands in the Atlantic, islands around Africa, islands in the Indies, islands in the Americas. The cluster of Spice Islands could seem as vast as Africa, the Canaries as massive as Italy. Europeans explored the globe mostly by sea and held their presence by ship; only in a few spectacular exceptions in Mexico and Peru were they successful on land, and then only through reinforcement by sea. When Coronado commenced his overland *entrada* into the Southwest in search of the Seven Cities, he sent a fleet up the Gulf of California in hopes of resupplying. The project failed on both counts, as did De Soto's, along with the Portuguese expeditions that plunged into Mozambique and elsewhere in search of opulent empires fancifully ripe for plunder and were never heard from again.

With so many islands there are many taxonomies possible. There were, for example, rediscovered islands and newly discovered. There were inhabited islands and uninhabited. There were real islands, tangible and mappable, and imaginary islands, either inherited from the past or conjured up out of fancy. Accounts and maps often contained them all.

Uninhabited islands were the easiest to manage, for they pitted capital and colonist against nature. Coming to one—for example, St. Helena—explorers would often deposit goats or seeds so they could reproduce and transform the island into a port for provisioning. If the island could support sugar, it invited settlement, either by free labor or slaves. Most offered ready-made fiefdoms by which to reward enterprising explorers.

Inhabited islands were both more promising and more problematic. Being inhabited did not solve the labor problem because it meant costly fighting, and when conquered, the indigenes often melted away under the blasts of social collapse and disease, or became slaves to be shipped elsewhere. The isles had to be repopulated. After the Guanches faded way, the Canaries were filled by Portuguese (80 percent). After the Taino collapsed, Hispaniola was staffed by African slaves, who introduced their old disease environment (think yellow fever) into their new setting. This was a different issue than transferring land taken from the Moors to a favored hidalgo. Still, inhabited isles were the models for settlement and governance that were applied to the sprawling lands of New Spain, which were many times larger than old Spain (Mexico alone was four times as big). Spain's New World holdings were over 20 times its size; Brazil was over 90 times vaster than Portugal. For seaborne empires, islands posed practical problems and served as prototypes.

And then there is the curious matter of imagined isles. They clotted early maps. Fantasy islands, from Atlantis to the Island of the Seven Cities, dappled the Atlantic. Some, like Antilles and Brasil, had their names transferred to discovered islands or mainlands. Some like the Seven Cities kept migrating into the still-blank areas of maps. Often they were the Sirens of discovery, calling gullible mariners onward. Erasing islands that did not exist was as big a task as penciling in those that did.

Other imagined isles flourished only in the mind. It was onto islands—self-contained sites untainted by contact—that Europe's intellectuals projected their social fantasies. The modern founder of the genre, Thomas More, published *Utopia* (literally, nowhere) in 1516. As a Platonic ideal, an updated *Republic*, Utopia did not have to be any place; but the rapid discovery of unknown lands made its literary location as a new-world isle plausible, along with the use of a weather-beaten Portuguese, Raphael Nonsenso, as a protagonist and prototype for the Ancient Mariners and Old Men of Belém who would hound and hector the enterprise.

From the monastic-styled Utopia of Thomas More to the bustling laboratories of Francis Bacon's New Atlantis, from the Brave New World of Shakespeare's *Tempest* to the tropical Edens portrayed by Pierre Poivre, the yet-undiscovered island was a cameo of cultural ideals. More's Utopia was modeled on a monastery; Bacon's Salomon's House, a factory for knowledge where "the End of our Foundation is the knowledge of Causes, and secret motions of things; and the enlarging of the bounds of Human Empire, to the effecting of all things possible." Echoing Homer, who had Odysseus narrate his tale from the isle of Scherie, Camões opens *The Lusíads* at Mozambique. But narration in medias res commanded less power than the prospect of new lands—new societies, new hopes—that could be lodged only on recently discovered or yet-unvisited islands. Columbus spoke for all such visionaries when he rhapsodized over Hispaniola—and for all scorners, when he failed to implant a society equal to that dream.[3]

Like imagined islands, utopian societies also migrated to an undiscovered Beyond as reality converted blank landscapes of hope into a messy palimpsest of lived realities. For every fanciful island deleted from the growing mappae mundi, another thrived in a world not yet encountered except in the human heart.

Islands, or isle-like surrogates, have persisted as a motif across all three ages of discovery. They served as navigational buoys, as signposts on the carreira da Índia, and as beacons of discovery. Pico de Teide on Tenerife became one of the indelible landmarks of the Great Voyages, not only for voyagers but for those chronicling what they did. Teide, and others like it, and mountains, typically volcanoes as were most of the isles, have remained beacons by which to navigate through the history of discovery that followed.

In the ages to come they would be revisited and recalibrated, and new isles, or isle equivalents, would be discovered. Hispaniola endured as an emblem of the human havoc caused by contact; St. Helena, of its environmental degradation. The Spice Isles that pulled Antonio de Abreu, Sebastian Rodriguez Cermeño, and Francisco Serrão to them became the place where Alfred Wallace conceived of evolution by natural selection; the Galápagos Islands gave Charles Darwin confirmation on his version of speciation. Tahiti was where Enlightenment science met Romantic sensibility. Easter Island became a melancholy parable of lost peoples. Tenerife was where Alexander von Humboldt field tested how to conduct scientific surveys of continents. New isles were found in submerged seamounts; new volcanic isle equivalents with black smokers in the deep oceans; new islands in space, with planetary moons.

And with extraterrestrial islands has come the prospect for otherworldly utopias. Gerard O'Neill sited his L-5 colony Dyson sphere at the Lagrange point between Earth and Moon, a floating Laputa beyond the atmosphere. Stanley Kubrick and Arthur C. Clarke's *2001* launched the future by means of portals between Earth's Moon and Saturn's Iapetus, through which astronaut Dave Bowman moves into another realm of being.

6

Portuguese Paradigm

I sing of the famous Portuguese
To whom both Mars and Neptune bowed. . . .
In Africa, they have coastal bases;
In Asia, no one disputes their power;
The New World already feels their ploughshare,
And if fresh worlds are found, they will be there.

—LUÍS VAZ DE CAMÕES, *THE LUSÍADS*

That Portugal pioneered the Great Voyages should alert us to exploration's uncertain origins and often desperate character. There was little in Portuguese history from which someone might predict, in 1450, that it would leap across whole seas and enter unknown continents; that it would expand its rivalry with Spain, itself a recent artifact of a marriage between Castile and Aragon, into the world's first global empires; and that tiny Portugal, perched on the land's end of a small, crenulated continent itself on the edge of the world's largest landmass would create the raw template for European expansion. Yet that is precisely what happened. For several hundred years, exploring nations sought to emulate the Portuguese paradigm or take over its imperial posts.

In almost every dimension of discovery Portugal led. It developed the critical nautical technologies—the caravel, cannon for naval warfare, basic navigational instruments and charts. It trained most of the pilots and many of the captains of the Great Voyages, even if they hailed from Italy or sailed for Spain, France, or England. Christopher Columbus, John Cabot, Ferdinand

Magellan, Amerigo Vespucci—all learned their craft in the Portuguese circuit and sailed for others only after Portugal had turned down their appeals. The navigators who discovered Newfoundland, Brazil, Africa south of the Canaries, the route to the Indies, China, and Japan—Diogo Gomes, Bartolomeu Dias, Vasco da Gama, Pedro Cabral, Francisco Serrão, Tomé Pires, Francisco Rodrigues—all were Portuguese. Azoreans were prominent in the Newfoundland fisheries; Madeirans, in the colonization of the Canaries. Afonso de Albuquerque showed how to turn discovery into empire. Francis Xavier showed how to turn it to missionizing. Luís Vaz de Camões showed how to turn it into literature. Within a generation, it came to be said that it was the fate of a Portuguese to be born in a small land but to have the whole world to die in.

What happened was that exploration became—directly, or indirectly through charters—an organ of the state, and because no single state dominated Europe, many joined the rush. Geographical exploration was a means of knowing, of creating commercial empires, of outmaneuvering political, economic, religious, and military competitors—it was war, diplomacy, proselytizing, scholarship, and trade by other means. For this reason, it could not cease. For every champion, there existed a handful of challengers. This competitive dynamic—embedded in a squabbling Europe's constitution—helps explain why European exploration did not crumble as quickly as it congealed.

It also became an expression of a people's culture. Portugal's overseas empire was a formidable commitment that drained perhaps a tenth of the population out of Portugal and compelled the country to defend what it quickly appreciated it could not. The flush of early wealth it acquired was soon destroyed by foreign wars that left it indebted, and ultimately in political hock to Spain. Others picked off its best Indies holdings. But while the interlopers poached pieces, a remarkably robust imperium endured. Portugal's saga told not of an endless quest that, once launched became unstoppable, but of a life cycle, one that birthed, grew, and then withered into senescence.

The Portuguese experience established the default setting for exploration's software. The degree of interpenetration between geographic discovery and Portuguese society was astonishing, of which a flotilla of exploring ships was only a down payment. All the founding explorers combined discovery with some other enterprise: affairs of state, commerce and conquest, prose-lytizing and poetry. Revealingly, all of them save Prince Henry and Luís Vaz de Camões died overseas.

Consider the gallery of founders.

Prince Henry was a late medieval prince, blurry-eyed speculator, and ardent crusader. But he was willing to look to the Atlantic—its isles, its African coast—as a site for colonization and trade, and an arena for com-petition with the Moors, and perhaps a bit of proselytizing. Critically, he fused exploration with state policy. There was little modern about him. Many of his discoveries were rediscoveries, and his purposes revived cru-sader chimeras. He did not embed the project into the constitution of Por-tugal, which meant that, when he died, discovery stalled. Yet he serves as a visionary, if a myopic one. He is the man who first sent ships on what became the carreira da Índia.

Vasco da Gama was the explorer as merchant, ambassador, and admin-istrator. Three times he sailed to India. The first from 1497 to 1499 blazed the passage and brought home hard-won knowledge about how (and how not) to navigate the Indian Ocean. From East Africa to Calicut he had the services of an Arab pilot and sailed with the winds; for the return, he ignored local lore and sailed against the monsoon winds, which nearly destroyed his fleet. He returned to Lisbon with one ship out of four and 55 men out of 170. But he came with a hold full of spices. He sailed back to India in 1502–3, after Cabral's epic voyage. Then he was dispatched again in 1524, this time to help clarify and clean up the fast-morphing mess that was Portugal's Estado da Índia. He died on the job, at Goa, perhaps poisoned.

Afonso de Albuquerque was the explorer as soldier and strategist. He sailed on two fleets to India, in 1503 and 1506, before returning as governor in 1509. Portugal early appreciated what the Estado da Índia required mil-itarily if a small country, for which it could take a hazardous year to send messages and men to its new empire, hoped to hold it against old and vastly larger rivals like the Arabs, Turks, and local rajahs. These strategic insights were codified in a 1505 *regimento* to governor Francisco de Almeida. Albuquerque confirmed and largely implemented them, quickly sizing up the military arena. He identified the critical straits that controlled passage into and out of the Indian Ocean and Spice Isles: the Red Sea, the Persian Gulf, the Strait of Malacca, the Mozambique channel. In a handful of years, from 1509 to 1515, he established the military basis for Portugal's presence. He attacked Turks, Arabs, and Persians, at Socotra and Hormuz and the Red Sea, and built fortifications to control those chokeholds. He forced Calicut into submission, built a fort at Diu, and recognized the value of Goa, which he made the administrative center of the Empire of the Indies. He took Malacca and opened the way to the Spice Isles, the South China Sea, and the Asian coast. He sent emissaries throughout the region, looking for allies, even among the Persians and Ethiopians. He anchored Portugal's seaborne empire with a circuit of fortresses. That Portugal went from discovery to rule was largely the achievement of Albuquerque, even as he suffered from court rivals in Lisbon, the suspicion of Manuel, and at one point an arrest. He was on the verge of being replaced when he died in sight of Goa.[1]

So much for gold and glory. God came later in the person of St. Francis Xavier. He was born a Basque in 1506, a year before Albuquerque first took Socotra and Hormuz. At the University of Paris he met Ignatius Loyola, a fellow Basque and a soldier who had undergone a profound religious conversion. They resolved to create an order pledged to poverty, celibacy, devotion to the salvation of believers and unbelievers alike, and a pilgrimage to the Holy Land. In 1534 they joined with five others to found the Society of Jesus, the Jesuits. Soon, they traveled far beyond the Holy Land to the realms of Iberian discovery. In 1540 Francis Xavier journeyed to Lisbon, and then to India, where he proselytized in south India among nonbelievers in numbers beyond his imagining. He then turned his attention to the Malay Archipelago, evangelizing at Malacca and among Dayak headhunters. In 1548 he returned to Goa, where more Jesuits had arrived and where a College of the Holy Faith was turned over to Jesuit care, at which native priests might be trained. He insisted that missionaries learn the language and adapt to the customs of the people they met. In Malacca he converted a Japanese man, Anjiro, and in 1549 sailed to Kagoshima to work further with what he considered "the best people yet discovered." Japan pointed him to China, and he was off its coast when he died in 1552. For the Jesuits he became the patron of missions; and Jesuits entered the ranks of explorers throughout the imperium of Catholic Europe. They were in North and South America, at the court of China, in India and throughout the Greater Indies, and in Africa from the Congo to the crumbling empire of Monomotapa.

What makes geographic exploration different from other forms of discovery is the journey, and no expedition is complete until it has its narrative. This, too, Portugal produced, not only in picaresque personal accounts but in the hands of Luís Vaz de Camões as a Homeric epic. Camões was born in 1524, during da Gama's last tour. He had a good education, but indiscretions exiled him from Lisbon, and after a tour in Morocco (where he lost his right eye), he returned, a changed man and at one point an imprisoned one. In 1553 he sailed for India, beat around the Empire of the East for 17 years, including Mozambique, India, Japan, the Moluccas, Malacca, and probably endured a shipwreck off the Mekong Delta before returning to Portugal in 1570. Two years later he published *The Lusíads*, a heroic poem celebrating the voyage of Vasco da Gama and his successors as they triumphed over nature and heathens. It took the classic epic as a model—this is the Renaissance after all. Camões was a half-blind Homer, and he adapted the inherited form (complete with Olympian gods) to celebrate the astonishing feats of his countrymen. The book soon established itself as the national epic of Portugal; his impact on Portuguese is similar to that of Cervantes on Castilian. He died in Lisbon in 1580, as Portugal was about to be absorbed by Spain.

In brief, Portugal did it all, and its Great Voyages transformed Portugal as little else could. They also destroyed it as few national undertakings might. Even as Camões penned his epic amalgam of triumph and tragedy, Pinto was recording a parallel, equally fabulous tale of adventure, mishap, and squandered opportunities that makes his *Perigrinição* read as though it were an early draft of *Gulliver's Travels*. The outcome depended on what else exploration bonded to. If a Portuguese had the whole world to die in, it was because Portugal had made that world possible.

Enthusiasm there was in 16th-century Portugal for expansion, but where and how and to what ends kindled a furious debate amid competing claims. Some wished to continue in and around the Atlantic; others looked always to the conquest of Morocco. Nine years passed after Bartholomeu Dias rounded the cape before da Gama headed to India, which was obviously not a self-evident thing to do, and may have been prompted largely to silence a small but noisy Indies cabal at the court.

Besides, the whole enterprise could wind down, as it did in slow decay within 60 years of its founding. The astonishing outrush lasted less than two generations, exhausted by overreach and especially by foreign wars (that Moroccan mirage). The colonies took on lives of their own, populated not by fresh émigrés from Portugal but by mixed-blood societies whose ties to the metropole were largely language and faith and ever more tenuous memories. The Empire of the East suffered from a deep rot of corruption and exhaustion. It had enemies everywhere, but the worst threats were internal. Its most profound wounds, as the Portuguese admitted, were self-inflicted. Yet a fierce pride persisted, as Charles Boxer aptly summarizes: "Disastrous as were many of their defeats by land and sea, humiliating as were the indignities to which they were sometimes subjected in places like Cacao and Madres, the Portuguese in the East were always proudly conscious of what they consider as their glorious sixteenth-century past." They were the descendants of Albuquerque and of a tradition that wealthier merchant companies could never attain and of victories and daring Asians could never match. "God, the Portuguese felt, was on their side in the long run, even if, as they frankly acknowledged, He was in the meantime punishing them for their sins by the loss of Malacca, Ceylon, Malabar, and Mombasa."[2]

In truth, more than an unanchored pride persisted. The ever-more-ramshackle empire endured; Angola and Mozambique did not become independent until 1974, Macau in 1999, East Timor in 2002. More significantly, the Portuguese experience established a template for exploration as something beyond foraging, piracy, and looting expeditions. It shone as an enterprise that could engage a society across its cultural spectrum and push beyond its old borders of understanding and expression. Not only the First Great Age of Discovery but those that followed built on a paradigm initially and impetuously inscribed by Portugal.

7

The Armada de Molucca Circumnavigates the Globe

From the time when we departed from that Bay until the present day we
had sailed fourteen thousand four hundred and sixty leagues, and completed
the circuit of the world from east to west.

—ANTONIO PIGAFETTA, A VENETIAN OBSERVER WITH THE
ARMADA DE MOLUCCA[1]

G reat Ages come with grand gestures. Within their chronicle of
travels, within their sprawling canvases of scenes, oversized personalities, desperate moments and fabulous triumphs, something
stands out not only as extraordinary but as exemplary that encapsulates
the means and ends of an era of discovery. It boils down the viscous sap
into a hard crystal. It speaks to what is best, most memorable, and most
significant—shorthand for centuries of messy voyages, treks, encounters,
sights, personal narratives, and serendipities that make up an age.

They are not the first bold announcements of discovery, which come early,
with the shock of first discovery or first expression from a Columbus, or a Dias,
or a da Gama. While that annunciatory first revelation can galvanize the
imagination and inspire successors, it comes too soon to capture the still-
inchoate features of the times, or is too tightly bound to a particular people
or project. Rather, grand gestures seem to follow 20–30 years later, after
exploration and its sustaining society have worked through the full terms of
engagement, as first discovery becomes full discovery. They are the moments

of exploring that more than any other capture the general imagination, that fuse place, time, discovery, and yearning in ways that seem to speak to an era's sense of itself. If they do not inform an age, they do display the vital attributes of an age as nothing else can.[2]

What might constitute the grand gesture of the First Age? Portugal could point to the carreira da Índia first launched under Vasco da Gama that completed the circuit of Africa, made the Estado da Índia possible, and became the basis for its national epic, *The Lusíads*. Spain might argue for the revelatory first voyage of Columbus, or perhaps Cortés's improbable conquest of Mexico, which anchored its empire in the Americas. England might look to John Cabot's daring sail to the New Found Lands; France, always willing to consider history pliable in the interest of *gloire*, to Jacques Cartier. But a grand gesture must transcend the scope of individual explorers or nations.

The essence of the First Age was, as J. H. Parry put it, the discovery of the sea, not simply the Ocean Sea that lay beyond the classical world or maybe encircled the lesser seas as the Styx encircled Hades, but the astonishing way in which they all connect and flow one into another. The enduring symbol of the age should highlight that fact. Surely, for the First Great Age of Discovery, this points to the circumnavigating voyage of the *Victoria*. Appropriately, its prime mover was a long-voyaging Portuguese captain, Ferdinand Magellan, a Portuguese serving Spain. Magellan was a paragon of his time, an explorer tenacious, religious, tough beyond reckoning and ambitious beyond yearning. God, gold, glory—all filled the mold of his soul.

The story of the Armada almost beggars the imagination. Five ships departed Seville on August 10, 1519, and after a shakedown cruise, the Canaries on October 3. Meanwhile, a letter warned Magellan that some of his captains planned to seize the fleet and that Portugal had dispatched a flotilla to intercept them, that, in effect, both Spain and Portugal regarded him as a traitor. When his belief that the Rio de la Plata was the strait he sought proved wrong, he stubbornly continued south, selecting Port St. Julian for winter quarters. On Easter night three Spanish captains mutinied. Magellan swiftly broke the rebellion. Then the *Santiago*, on a scouting mission, sank. The remaining fleet probed south. In late October, two ships were driven by yet another storm into a bay, and so by accident discovered the passage they sought. Several days later, passing between mountains, they entered what

Magellan named Mar Pacifico. The captain of the *San Antonio*, the armada's supply ship, then deserted, wound back through the strait, and returned to Spain with an alternate version of events that cast Magellan as the villain. The long crossing over the Pacific—109 days—was maddening, the crew reduced to eating rats, and sawdust, and oxen leather soaked in seawater. On March 6, 1520, the three ships reached Guam, and ten days later, the Philippines. Here Magellan, for the first time, lost control, not only of his crew, but seemingly of himself. He practiced faith healing, tried mass conversion, and then, to demonstrate the power of Spanish arms, arranged a pointless battle at Mactan during which he was slain in the surf. Having been shown not the Europeans' invincibility but their frailty, their Philippine hosts betrayed and slaughtered 30 crewmen, mostly officers. Six months later, the *Trinidad* and *Victoria* reached the Moluccas. The *Victoria* departed to the west, effectively entering the Portuguese Indies. The damaged *Trinidad* stayed for repairs, then turned east, where it was intercepted by the Portuguese; its Portuguese crewmen were executed; and the enfeebled ship itself sank in port during a storm. The *Victoria*, now captained by Juan Sebastián Elcano, successfully threaded its way across the Indian Ocean and back to Spain, arriving on September 6, 1522. One ship out of five, and 18 men out of the original complement of 270 returned.

It's a saga that could only have happened amid the Great Voyages, and it crystallized as nothing else in the era could the complex means and motives that powered the age. In its captain's unyielding vision, a compound of delusion and determination; in its amalgam of treachery, betrayal, and undaunted fortitude; in its confused, ultimately lethal mingling of Cross and sword; in the sordid competitions it displayed, both within the armada and between it and rivals; in its appalling costs and world-altering triumph—the Armada de Molucca easily qualifies as the grand gesture of the First Age and a template that taunted later ages to emulate. It confirmed for the first time that it was possible to put to sea anywhere in the world and end up anyplace else. The act was utterly unprecedented. It fired the imagination of its time. In its scope, audacity, and ferocious display of will, it dazzles still.

8

Encountering

These barbarous people, who I dare say judge all other nations by
themselves, would not venture close to us. . . . They behaved rather as we
behave towards victims of the plague: our people were obliged to take the
goods which they wanted to exchange for fish some distance from the ship
and then turn back. After the natives had observed this, they approached,
took what had been brought to them, put their fish in the same place, and
returned to their huts.

—C. JANNEQUIN, OFF THE COAST OF GUINEA (1640)[1]

When Columbus made landfall in the Antilles, he followed what
had already become a protocol for new discoveries. He looked
first to the safety of his ships—wait for dawn's light, watch for
shoals and reefs, find a protected harbor or lee anchorage. Then the admiral and captains would take the longboat ashore, kneel, and give thanks.
As soon as possible they would try to locate natives (who typically were
already watching the newcomers, and often came to the ship in boats), display emblems of authority, communicate by signs and perhaps share gifts,
begin to train interpreters, and collect directions to headmen, and, always,
identify the way to wealth. Amid so many isles, the crews had ample time to
drill and adapt the protocol, but behind them stood more than seventy years
of Portuguese probing among Atlantic isles and African coastlines.[2]

Nature threw plenty of hazards at the explorers, especially when it
was nature encountered for the first time. Unexpected storms sank fleets;
unknown shoals wrecked ships; doldrums becalmed mariners for weeks.
Spaniards had never heard of hurricanes until they settled in the Caribbean

(the term comes from Taino), and then never forgot them. But the recurring hazards were social. Survival depended less on hunting, wrestling with grizzlies, and staggering through sandstorms than with cross-cultural politics and negotiations. The failure to deal with indigenes could ruin an expedition, or destroy any hope for a return visit. The inability to handle crews could unhinge an armada. Even many natural risks resulted from the failure to learn from locals or the failure to cope with potential mutineers; and always there were prickly monarchs, suspicious of would-be Caesars across the sea and distracted by the temptations of rivals.

Uninhabited lands had advantages in that they did not require negotiations, wars, or treaties; they were literally there for the taking. The Atlantic isles, apart from the Canaries, are good illustrations. But most landfalls occurred where people existed. Discovery meant contact, which meant an encounter, and encounters could be multidimensional, mutual, and messy because they existed not simply in physical settings but within cultural geographies. An encounter could mean negotiations, possibly fighting, problems of understanding, ethical rights and responsibilities, and issues of governance. It posed questions of identity and the nature of the world and how it might be comprehended. Encounters with other peoples powered the moral drama of discovery.

It was rarely a simple transaction. Europeans voyaged for wealth and trade, which required someone to trade with who had something Europeans wanted and who wanted something Europeans had. The zamorin of Calicut laughed at da Gama's sample goods. The Chinese considered the newcomers as unwashed barbarians and found nothing worth trading for except bullion. In its minimalist form, mariners would deposit goods on a beach, and return the next day, while locals would take the goods and leave in return what they regarded as a fair exchange. Locals might trade gold, ivory, or fur for wool or iron tools. If recurring trade was possible, there might be agreements or even treaties. If the wealth was great, or if the other party balked, conquistadores might choose to seize it. Fortune remained the prime mover of expeditions, the motive force to which all other gears connected. Voyages had to pay for themselves. Princes and nations sailed to increase their wealth, power, and prestige.

Economics, however, could not be disentangled from social relationships, military interests, politics, and religion. To learn about local circumstances, explorers relied on native guides and translators. To protect trade Europeans frequently found it necessary to intervene in local politics, and to seek military alliances, often exploiting a divide-and-conquer strategy. To help sanction what could be a brutal project, discoverers held the cross in one hand and the sword in the other, proselytizing to the true faith; and monks and priests could become explorers on their own, searching out new souls as conquistadores did caches of bullion. To maintain a sufficient population, settlers imported slaves and interbred with locals, creating mestizo societies. Only in a few instances, and then outside the First Age, did immigration from Europe occur on a large scale. Until then trade dictated small settlements for provisioning, not wholesale demographic takeovers. Large colonies came with big costs; after a first flush of looting, they typically became burdens. Exploring nations sought to increase their wealth, not spin off versions of themselves.

Even in the 15th century enthusiasts regarded the Canary Islands as a crucible. They were, for Portugal, an island too far and for Spain, a port of departure for empire. They had a bit of everything and mostly had it first. Their story was recapitulated in Hispaniola, which received the fleets the Canaries dispatched. They are the template for a great set piece of exploration, the encounter.

Their rediscovery led to conquest, exploitation (including slaving), settlement and plantations, missionizing, and ultimately, because they were isolated islands, the extinction of the indigenes—a cameo for what contact could mean. The reduction of the islands took nearly a century, from 1402, when Jean de Béthencourt and Gadifer de La Salle sailed to Lanzarote and Fuerteventura in search of gold, until Spain finally subjugated Gran Canaria in 1483 and Tenerife in 1496. The reason for the slow conquest is primarily the indigenous Guanches; they were better equipped to defend their islands than Portuguese and Spanish conquistadores were to seize them. Decade by decade attrition by war, slaving, and co-option took its toll, but it appears that in 1481 the newcomers introduced a plague (*la peste*) into Gran Canaria and in 1495 to Tenerife (*la modorra)* that reduced Guanches' numbers and broke their capacity to resist.[3]

"Pacification" was a long slog and a prolonged period of contact, during which European organisms remade the landscape, Europeans and their slaves repopulated the islands, and European scholars and rulers debated the character of the indigenes and what rules of governance might apply. All this foreshadowed what was to come. There was even a miraculous manifestation of the Virgin Mary to a Guanche shepherd, heralding similar appearances throughout the Spanish New World.

The Canaries were a prototype, but many encounters assumed forms that played with its parameters. Consider as exemplars three classic expressions from Portugal's expansion: Elmina, a trading post; Goa, an administrative capital for empire; and Kongo, a missionary adventure.

São Jorge da Mina (Elmina, "the mine" due to its access to gold) was a permanent, as opposed to seasonal, trading factory erected at Cape Coast, halfway across the Gulf of Benin. It was ideally placed to intercept the trade in gold and ivory that had previously trekked north across the Sahara. Begun in 1482 it epitomized the Portuguese dream of capturing the African traffic at its source. It could resist both Castilian interlopers and local chiefs, and could extract better prices from locals than could ships that appeared seasonally and could not wait out negotiations. The *feitoria* was a complement to the islands that framed the gulf—the Cape Verde archipelago to the west; Fernando Pó and São Tomé to the east. It was sited on grounds that made it almost an island itself, a moated castle. There was no mass settlement, no expansion of rule into the interior: this was a trading post that left to the locals the task of assembling suitable goods. Almost immediately it thrived. For the next 500 years it continued, changing hands as one European rival after another claimed it and its commercial hinterland.[4]

Goa was much more: it was not only a center for trade but for outright rule, the keystone block for a ring of fortresses that could control trade in the Indian Ocean. It was the administrative capital for the Estado da Índia. It was not an integral part of the spice trade, but it was for the trade in horses (horses could not breed in south India and had to be imported), and so was vital for military purposes. That role continued under Portugal. Goa was taken by force and held by force; it existed to apply force to ensure the survival of the empire.

Shortly after Albuquerque became governor, the sultan of Bijapur died, leaving Goa, a major port city, undefended. Goa was already in Portugal's sights; Albuquerque expanded his brief not just to reinforce the fight against Gujarat, Calicut, and Egypt, then ruled by the Turks, but to establish Goa and its hinterland as Portuguese; not just another fortified feitoria but the capital of a colony. It would serve as a safe port for provisioning; a military post for enforcing rule, building ships, and housing munitions; and a sanctuary for a permanent Portuguese presence (it eventually housed the Jesuit college for training local clergy). The city rose on a delta, almost moated, but easily accessible from the sea. With an amphibious landing in February 1510 Albuquerque quickly seized the city. When he left for other tasks, the new sultan retook the city with the help of Turks and proceeded to stiffen its defenses with artillery. Albuquerque returned, this time for a vicious, risky Reconquista that lasted until December, at the end of which he massacred the garrison and married off surviving women to his soldiers. This was the Portuguese equivalent to Cortés reconquering Tenochtitlán, and it can stand in cameo for so many other Portuguese colonies populated by intermarriage with locals. With Goa secure Albuquerque turned to Malacca, which he took with 800 men against a large army, heavily bolstered with artillery and even war elephants. Internal dissension among the Malaccans helped. So did the character of Portuguese soldiers, fighting like berserkers against enormous odds. Malacca was the Portuguese answer to Spain's sack of Aztec and Incan empires. With Malacca in his hands, Albuquerque dispatched expeditions to seek the Spice Islands themselves.[5]

The Kongo tells a very different narrative. In 1482, the year construction began on Elmina, Diogo Cão beat against coastal currents into the mouth of the Congo River. Initial contacts seemed friendly and mutually useful, and an exchange commenced of visitors as well as goods. Emissaries from Kongo sailed to Lisbon; craftsmen and missionaries sailed in return. The Kongoese seemed to consider Portuguese contact as a full-spectrum encounter, not just an opportunity to exchange trade goods and acquire new weaponry but to absorb the architecture, norms, and beliefs that bolstered Portugal's economy and military prowess. The Portuguese were allowed to erect a church; missionaries followed. In 1491 King Nzinga Nkuwu was baptized. Later victories over rivals he attributed to interceding visions of the Virgin Mary and St. James of Compostela. "Against all the odds and against all expectations," Felipe Fernández-Armesto observes, Christianity flourished. It spread first in the court, then, often by coercion, into the countryside. It acquired further political power as a "spirit cult exclusive to the king"; and when a successional fight was averted among the king's numerous offspring from his many wives, Christianity—which promoted only one wife, the first—came to the aid of the first wife's first son, also baptized, who took the name Afonso. His own son, Henrique, he sent to Rome and had consecrated as a priest and bishop.[6]

But however much the Cross could absolve the sins of contact, however loudly for public consumption the salvation of souls was broadcast as cover for conquest, gold (by which "all things are possible," as Columbus put it) still trumped God. Lisbon sent more traders and soldiers than missionaries. Unfortunately, Kongo had little of the gold and ivory that Portugal lusted after. What it had in abundance were souls, who could also be slaves. A backlash developed, as Portuguese slavers, in particular, stirred up anger by poaching on royal monopolies; Portuguese racial castes poisoned social relations; and Portuguese hypocrisy overcame Kongoese infatuation. In the face of Portuguese indifference, the Christianization of Kongo faded.

In that it can stand for other missionary efforts throughout the Estado da Índia. The apostolic St. Francis Xavier had baptized thousands in his wanderings from Kerala, India, to Malacca to Japan. The secret was not just to convert from the top, but to immerse oneself in the local circumstances, learn the native languages, behave in ways that spoke to local traditions, and create an indigenous clergy. This is the same formula for successful settlement and trade other than by the sword. In some places the new religion grafted or rooted; in others, it withered or was suppressed. The 16th century was Japan's "Christian century." Jesuits made it to the court of the Ming dynasty in China, replacing literate monks who had traveled the length of the Mongol empire, even serving as astronomers and helping meld Chinese and Western mappae mundi. Then, like so much of the Portuguese tsunami that washed over the shores of the east, most of Christianity's presence receded along with the surge waters generally. It endured in a few sites of sturdy settlement, especially among mestizo societies. These became living versions of the padrãoes deposited along the carreira da Índia—claims not always fully settled.

The Cross added to discovery, supplemented the exchange of knowledge, and in the end stumbled along with so much of the implausible empire. For missionaries the momentum passed to the Spanish and later French empires in the Americas, where their achievements acquired the aura of an epic. But they were promoted, or tolerated, so long as they advanced the interests of the state. When the Jesuits became pesky, they were expelled—from Portugal and its empire in 1759, from France in 1764, and from the Spanish empire in 1767. By then secularism, which passed across the face of the religious West like the terminator across Earth, was replacing missionaries with naturalists, and religious knowledge had to be recoded to match the algorithms of modern science. Missionary explorers survived by collecting plants and butterflies rather than souls.

Of the whole cabinet of curiosities that holds the diversity of encounters, a special place has been reserved for the original. First contact was, potentially, exploration stripped of the qualms, morasses, and banalities that were sure to follow as discovery morphed into commerce, contact into imperialism, and new knowledge into scholasticism. Here, physical adventure held the possibility to undergo a transubstantiation into moral drama. The best first contact could change both parties. It could force a reevaluation of the world and our place in it.

That, at least, is how contemporary advocates for space exploration like Carl Sagan have presented the case. In truth, anything approximating that scenario rarely happened. Contact was often mundane, any surprise soon dissipated, and rather than shake apart one side's cosmology, the old order adapted to accommodate the new revelations. Besides, written records tell one side and come with their own conventions. It is very hard to know what actually happened and how each party interpreted it.

But there is one extraordinary exception because it occurred in recent times, and that is the saga of the gold-seeking Leahy brothers when they trekked into the unknown interior of New Guinea in 1930. Not only were they seeking new lands, but they did so as self-conscious explorers, and they carried a motion-picture camera to document what they saw and experienced. When they passed over the summit of the Bismarck Mountains, they found the holy grail of discovery: a new land and an unrecorded people.

They looked down on immense highland valleys, densely cultivated, flush with fires, awash with people unknown to them and who themselves knew nothing of the world beyond their tribal borders. The prospecting expeditions, under Mick Leahy, continued into the mid-1930s, extending discovery into valley upon valley, beyond the vales of Goroka and Asaro, and meeting tribe after tribe, pushing on to the putative source of the gold, much as Hernán Cortés kept moving inland until he found the great depository of Aztec wealth. That first expedition ended up crossing the island, a veritable microcontinent. Here was raw first contact of a sort not seen since the Great Voyages and the entradas of the New World conquistadores.[7]

In truth, it is unsettling how fully the Leahys' encounters echo those of previous centuries. Without a common language, exchanges were limited to signs, pantomimes, and demonstrations. The Australians wanted food, information, safe passage, and, later, women and workers; the highlanders wanted shells (their equivalent to bullion), steel axes, and, later, weapons. The explorers demonstrated their firepower and their superior technologies, and in later years sent a few select indigenes by airplane to see coastal cities; they took youngsters who could learn their language and serve as interpreters. The highlanders sought to fit the strangers into their existing economic and political dynamics as well as their prevailing cosmology; they wanted the unbelievable wealth of shells the strangers could disburse, and they sought to exploit the newcomers to advantage in the complex balance of endless wars among neighbors. Quickly, the Leahy party evolved protocols for contact.

Yet the encounter was profoundly asymmetrical. The explorers had come without permission or prior notice. They had simply appeared and, by their sheer presence, broke the old order. They stood outside the existing etiquette of exchange, were not subject to taboos, did not warrant traditional courtesies to travelers or pilgrims, were neither friend nor foe, just a pale Other. Their very identity was upsetting. Given the options, the highlanders labeled the newcomers as the spirits of former tribal members now returned to life. So, too, they sought to incorporate the cornucopia of shell wealth and the weaponry of the interlopers within their existing economic and political contexts, Unsurprisingly, where communication was limited to crude signs and barter, violence was almost inevitable. The explorers were determined—believed it essential to their survival—that they demonstrate their lethal weapons. The indigenes wanted, first, to seize the wealth of the intruders and then, once they understood the folly of direct force, to steer that violence against their hereditary enemies.

In the single-mindedness of their gold lust; in claiming special spiritual powers by predicting events from eclipses to the arrival of a Junkers transport plane; in reliance on indigenous labor and local lore; in their sexual relations with native women; in their violent retaliations to theft; in the cultivation of interpreters and a common pidgin language of trade and travel; in the white-hot rivalries occasioned by competing prospectors and the bitter quarrels over priority that followed; in their awkward relations with missionaries and colonial authorities; in their destabilizing presence; in their growing weariness over the endless violence around them and the sheer strangeness of an Otherly morality; in an exhaustion that could lead to either submersion into that order or a desire to exterminate it—the Leahys were a throwback to the First Age. They recapitulated a historical scenario of the civilization they represented. They demonstrate in modern garb what early encounters were probably like.

Each group was a novelty to the other. But the Leahys coped more easily. Why? They knew what they wanted, where they wished to go, and how they proposed to get what they sought. The New Guineans did not. The Leahys were surprised, but not stunned. They carried in their baggage train half a millennium of cultural tradition based on encountering new peoples in new lands. The New Guineans knew no one outside their hostile neighbors, all of whom worked to keep one another strictly in their place. They learned quickly enough; but the momentum lay with the explorers, as it had so often in the past. It was when the explorer returned that troubles so often boiled over. That had happened with Columbus and Cortés, and it happened with the Leahys. Familiarity bred not only contempt but covetousness, as each party sought to turn the other's strengths to its own advantage. The vaunted impact of first contact only occurred if it survived the more complex contacts to come.

Encounter was the great set piece of exploration that kneaded character and conflict into narrative. It was less a simple process of reaching a goal—the source of the Nile, a passage to India—than it was an exchange between people whose aftershocks could affect the discoverer as much as the discovered. It was, not unlike Renaissance humanism, an exercise in translation and interpretation.

Throughout, discovery was mutual. Europe's discovery of sub-Saharan Africa was also Africa's discovery of Europe, Christendom's encounter with Muslims was also a Muslim encounter with Christendom, Europe's engagement with Amerindians was also an Amerindian engagement with Europe. Indigenes might be as shocked by European norms and practices such as monogamy and forced conversion as Europeans were with cannibalism, polygamy, and human sacrifice. Each group could be cruel according to its own standards. Each could find nobility in the other. Each traded and maneuvered to its own advantage. Encounters were less a series of straight transactions than complex chess matches in which understanding the other was vital to success. Europeans might try to impress the natives by taking them to court, but courtiers might be more influenced by the natives than the natives by them.

So while Columbus might exalt, "How easy it would be to convert these people and make them work for us," the reality was that conversion might go either way. It did not always end with the triumphalism of the discoverers: it could cause the thoughtful to question their own assumptions, values, sense of their place and purpose in the world. However culturally corrosive contact might be for those Europe encountered, it could be equally corrosive for Europeans. The threat was present, at both an individual and civilizational level.

It began early, as Cortés discovered at Cozumel. Through an Indian interpreter named Melchior (who "understood a little Spanish and knew the language of Cozumel very well") Cortés learned of two Spaniards who were held as "slaves" by Indians farther inland. One, Gerónimo de Aguilar, was freed, while the other, Gonzalo Guerrero, chose to remain. Aguilar told their story. They were the refugees of a voyage from Darien to Santo Domingo, brought on the wrecked ship's longboat by currents to Campeche. Some fifteen men and two women had escaped the downed ship. Upon the refugees' landing, the indigenes they encountered sacrificed some "to their idols"; others died of disease; the two women perished "of overwork"; but Aguilar and Guerrero had escaped, and now they alone survived. So far this was almost a parody of those narratives in which the indigenes found themselves under Spanish control. The climax came when the captives had to choose which society to follow. Among the first of New World captivity narratives, this one was deeply ambiguous and ironic.[8]

When Aguilar appeared, he squatted "in Indian fashion" and wore no more clothing than the natives. He had taken holy orders, was grateful to be rescued and reclaimed by Spain, and willingly accepted clothes and relearned Spanish. But Gonzalo Guerrero, a sailor from Palos, did not wish to return to Spain. Tattooed and pierced, he was married now, a father; a cacique and a "captain in time of war," he had been absorbed into indigenous society. Aguilar reminded him that he was a Christian and "should not destroy his soul for the sake of an Indian woman," and if necessary could take his family with him to Spanish settlements. Gonzalo refused; and "neither words nor warning" could persuade him otherwise. He had converted.

He was not alone. As accounts of encounters mounted, as natives were brought willingly or not to Europe, intellectuals could experience the same questioning. Surely the most famous is Michel de Montaigne's 1580 meditation on a Brazilian indigene, identified as a cannibal. Montaigne begins, as a good Renaissance humanist would, with a review of the literature. He cites Plutarch, Plato, Virgil, Horace, Aristotle, Lycurgus, Seneca, Suetonius, Herodotus, Chrysippus, Zeno, and Juvenal for comments, quotes, and examples.

Against them stand the witnesses from contemporary voyages—testimonies from the book of nature, not texts from antiquity, and so many, tumbling over themselves, foaming like a stream over the boulders of ancient manuscripts. Better, there were three Brazilians, "ignorant of the price they will pay some day, in loss of repose and happiness, for gaining knowledge of the corruptions of this side of the ocean; ignorant also of the fact that of this intercourse will come their ruin."[9]

He spoke to one of them "a great while," though through a poor interpreter—translating again, so vital to both humanism and exploration. The Brazilian was not awed by the pomp of the court or the technologies of which Europeans were so proud so much as he was baffled by a society that let itself be ruled by a child and that knew such chasms between wealth and poverty. There lay the critique and the method: to see themselves from the outside, viewing themselves as they would newly discovered peoples, to refract criticism through the reflected image of a looking-glass Other. It was a universal issue, but one made urgent for Europe by its pursuit of exploration.

Montaigne concluded that "there is nothing barbarous and savage in that nation, from what I have been told, except that each man calls barbarism whatever is not his own practice." We have, it seems, "no other test of truth and reason than the example and pattern of the opinions and customs of the country we live in." Where we grow up is "always the perfect religion, the perfect government, the perfect and accomplished manners in all things." We call wild what nature has produced, and civilized what artifice has shaped, but we might as correctly honor as civilized what remains close to nature and condemn as wild what has strayed. ("Truly here are real savages by our standards; for either they must be thoroughly so, or we must be.") Of the two Montaigne would prefer a Brazilian cannibal to a Versailles courtier. The nominal savage is actually noble by nature, living in a state of natural law, "purity," and innocence, surpassing "not only all the pictures in which poets have idealized the golden age and all their inventions in imagining a happy state of man, but also the conceptions and the very desire of philosophy." The evidence of exploration has produced in the flesh what Europe's theorists had only conjured in the mind.[10]

So was birthed the Noble Savage, and eventually a line of reasoning that would lead by the 20th century to a concept of the cultural relativity of all peoples. While Europe's discoveries provoked some of its intelligentsia to become apologists and panegyrists for imperial rule, they prompted others to reverse relationships and condemn Europeans as the real savages. Cultural relativism spawned moral relativism, anathema equally to the religious zeal and the lust for gold and fame that animated the First Age. The acceptance of cultural relativism required more than the Great Voyages, but it is difficult to see how it could have occurred without the global encounters those travels set into motion, or what other civilization might have come to such a conclusion.

But the challenge could be even more fundamental. Discovery did not simply look out. It also looked back at the images reflected from others. It compelled Europe to ponder not only its own creeds and precepts but the very act of restlessly seeking out the new. "Poor wretches," Montaigne fretted of his Brazilians, "to let themselves be tricked by the desire for new things, and to have left the serenity of their own sky to come and see ours!" That, too, might be reflected back onto those Europeans whose fecklessness and lusts had shoved and yanked them and their whole civilization to new lands. "We grasp at all but catch nothing but wind," he lamented. Yet those winds filled the sails of far-ranging marinheiros.[11]

What the explorer often acquired—what the sensitive observer like Montaigne could not help but have—whether pinning it like a beetle to a collection box or getting it into his blood like malaria—was an appreciation that could evolve into an empathy for alternative cosmologies, for understanding and often admiring the perspectives and codes of his native companions or their representatives even though that appreciation might be profoundly unsettling to his own culture's existing order. The very nature of contact made such an exchange inescapable. If the discovered people had not wanted to be discovered, neither did the discovering people wish to see their own beliefs frayed and unwound. Yet that is what, inevitably, exploration did.

9

The Other New World

Your greater navigators will unfold
New worlds to the amazement of the old.

−LUÍS VAZ DE CAMÕES, *THE LUSÍADS*

There were, in truth, two new worlds being discovered. One was geographic, the other, intellectual. The discovery of new lands coincided with the recovery of ancient texts. The Great Voyages sailed in parallel, a loose valence, with the Renaissance. What happened with Ptolemy's map happened with Herodotus' ethnography and Aristotle's physics—they were found wanting, and eventually discarded as relics. But while the Renaissance incubated in Italy and then spread selectively to Europe, the Reconnaissance surged out of Iberia and spread across the globe.

Its empirical discoveries spilled beyond the borders of formal learning, and then became a metaphor for discovery in all fields. Andreas Vesalius opening up cadavers and challenging Galen on medicine was of a piece with Amerigo Vespucci questioning Ptolemy. Michel de Montaigne's essay on cannibalism opens with a litany of quotes from ancient authors, then confronts a real Brazilian native (from "Antarctic France"); the essay is itself a test, an assay, as its genre name states. In 1620 Francis Bacon opened his paean to a new age of learning, *The Great Instauration*, with a frontispiece that shows two vessels sailing beyond the Pillars of Hercules, signifying the limits of the ancient world. There was a new world of natural philosophy, what would eventually become known as science, waiting to be discovered by adventurers willing to sail beyond the borders of old scholasticisms. When he imagined a new world based on systematic experimentation, Bacon placed his New Atlantis, as Thomas More had its monastic Utopia, among the yet-undiscovered isles. By this time (1627), however, the Atlantic had no unvisited places, and Bacon had to site the isle, Bensalem, in the still-mysterious Pacific.

Exploration proceeded in couplets. Islands were inhabited and uninhabited, mainlands were known or new, people were primitive or civilized, Christian or Moor; discovery paired with rediscovery. Except for those scattered sites that held no people, it was mostly a process that transferred knowledge from indigenous peoples to explorers. As with Renaissance humanism, the basis for exploration was translation, and as with Renaissance scholarship, Europe wanted to know more from others, much as it sought their goods, than they wanted to know about Europe or valued what it had. Europe had more to learn than to teach.

What, after all, did it mean to "discover" India or China or Ethiopia? All were known, if poorly—they were, after all, the objects of the entradas and voyages. The essence of the mythology of first contact is the shock of novelty, of the unexpected. But what does first contact mean when the unexpected leads to what is already known? That is what happened with da Gama's fleet as it anchored off Calicut. On shore for a quick survey João Nunes was greeted and directed to some other merchants then in the city. "May the Devil take thee," they exclaimed in Castilian. They were Tunisians who had traveled the Muslim trade routes to India. Recording the incident, Álvaro Velho wrote with measured understatement that "we never expected to hear our language spoken so far away from Portugal." They had traveled around the ends of the world only to meet those they had sought to escape.[1]

Language, translation, interpretation, and glosses—these were for the scholars of discovery the critical skills. But even explorers knew the vital need to communicate with the natives. Much as they would deposit goats on discovered isles as a source of future provisioning, so they collected guides and translators, and gathered youngsters who could naturally learn the new languages and assist in communication. In the first island he made landfall

Columbus "took by force" some indigenes "in order that they might learn and give me information of that which there is in those parts, and so it was that they soon understood us, and we them, either by speech or signs, and they have been very serviceable." It helped that there were those "who navigate all those seas, so that it is amazing how good an account they give of everything." Magellan had a stroke of fabulous fortune when his servant, Enrique, whom he had acquired years before in Malacca, was a Filipino who could speak to the natives of Cebu. On his own circumnavigation Thomas Cavendish had with his crew a Spanish navigator, a Portuguese veteran who had been to China, two Japanese mariners, and three boys from the Philippines. Bernal Díaz explained that "without Doña Marina," who spoke Tabascan and Coatzacoalcos, "we could not have understood the language of New Spain and Mexico" and the conquest would have faltered. When explorers returned, their own discoveries became known through translation and printing, even when they brought indigenes back with them. Montaigne lamented that, although talking with a Brazilian for a long time, the interpreter "followed my meaning so badly" and stupidly "that I could get hardly any satisfaction." The Great Voyages were as much about languages as about rigging and sails.[2]

Discovery could mean very different things when exploring the Cape Verde Islands, south India, and Newfoundland. It could mean that the existing stock of knowledge was revised, or it could mean that explorers now knew something they hadn't even imagined before.

In *The Wealth of Nations* Adam Smith declared that the discovery of the New World and of a passage around the Cape of Good Hope to the Indies were the two greatest events in world history. But those revelations did not immediately upend existing maps and worldviews. The Renaissance spread of its own momentum without afterburners from exploration; the impending Reformation did more to remake European civilization than did its early explorations. Doubts about Ptolemy and other inherited perspectives were already in the air, and the first triumphs of the Great Voyages only enlarged, not revolutionized, them. Columbus discovered islands—this was to be expected and what he had claimed he would do. Dias and da Gama found a sea route to the Indies—this was the point of Portuguese exploration, and it led to where Portugal wanted to go. There was confusion over where exactly those Columbian islands were, whether in the Ocean Sea or offshore to the Indies. But scholars slotted the initial information into existing categories and expectations, much as officials adapted institutions devised for Madeira and the Canaries to St. Helena and Hispaniola. The same held for discovered plants, animals, and peoples.[3]

The idea of a New World intervening between Europe and Asia came slowly and fitfully. The phrases *otro mundo* (other world) and *novus orbis* (new world) were bandied about, but still referred to new lands, in particular to islands, not to continents as large as Europe and Africa combined. The magnitude of the discovered places did not become apparent until Amerigo Vespucci returned from his voyages and a popular account, *Mundus Novus*, was published in 1501. Martin Waldseemüller fixed the realization into his massive 1507 compendium, *Cosmographiae Introductio* and its mappae mundi and labeled it "America," playing off Vespucci's name. By 1511 the discoveries had gone far enough that even their bustling chronicler, Peter Martyr (who confessed that he wrote down "everything in haste and almost in confusion, as the opportunity offers, and I cannot observe order in these things because they happen without order"), called America an *orbe novo*. The New World it has remained ever since.[4]

What was true for geography proved equally true for all the other discoveries that the ships hauled back in their holds, that chroniclers listed, and that artists tried to render. The new animals—bison, llamas, alligators, rattlesnakes. The new plants—tobacco, pineapples, tomatoes, potatoes. The new peoples—the Beothuks, the Tainos, the Aztecs. Most such novelties came by translation from indigenes, some were imposed by the necessities of conquest, and some out of the same wide-eyed mix of curiosity and cupidity that had brought the newcomers to far lands in the first place. The quest for new wonders could at times match the pursuit of gold. Here is Bernal Díaz interrupting his narrative of the march to Mexico's capital by the encounter with a "volcano near Huexotzinco, which was throwing out more fire than usual." All of us, "including our Captain, were greatly astonished at this, since we had never seen a volcano before." One of the captains,

Diego de Ordaz, asked for permission to "see what it was"; and Cortés then "expressly ordered him to make the ascent." He took two other Spaniards and some local Indian chiefs, who "frightened him with the information that half way up Popocatepetl—for this was the volcano's name—the earth-tremors and the flames, stones, and ashes that were thrown out of the mountain were more than a man could bear. They said the guides would not dare to climb further than the *cues* of those idols that are called the *Teules* of Popocatepetl." While the Indians paused, the Spaniards pressed on. Ordaz later received permission to include the volcano in his coat of arms. Then the army returned to its march and the even grander marvels of Tenochtitlán.[5]

Early discoveries predated modern science, and reports took the form of letters and chronicles—this is, after all, a profoundly text-based scholarship. Scholars then assembled and analyzed the data in the manner of their training, which is to say, the emerging genres of Renaissance humanism, themselves based on classical literature. Studies took the form of herbals and catalogs of useful flora and fauna, and of course of surprising novelties (some real like jaguars, some fanciful like Patagonian giants). The centers of discovery were rarely the centers of dissemination. The major cartography was done in Germany and the Netherlands; the primary node for circulating news was Italy. Quite apart from the printing press, existing networks of correspondence spread the word quickly.

How discovery entered formal lore is nicely illustrated by Spain. Peter Martyr (by birth an Italian) served as *cronista* to the Council of the Indies and chaplain to Queen Isabella, which kept him close to court news and gossip. He maintained a wide community of correspondents, mostly in Italy. In 1493, beginning with Columbus's return, he began to keep a detailed chronology and gloss on discoveries; from 1511 to 1530 he organized that record into ten volumes that he called the *Decades* (*Decas*); these were summarized in 1530 as *De Orbe Novo*. Paradoxically, commentators like Martyr could see what explorers like Columbus could not. What humanists did by gathering dispersed texts, Martyr did with the fresh texts of discovery. The academics could be wrong—often were because of their deference to the ancients. But they also helped to spread credible news and kept as authentic a record as possible.

Most, like Martyr, never left their study or court and handled only interviews and documents. But some were on the scene—were present and participated. Fernández de Oviedo y Valdéz, for example, arrived at Darien in 1514, published a short survey (*Sumario de la Natural Historia de las Indias*) in 1526, was appointed the official chronicler for Spanish America, and then, amassing data and collections painstakingly over many years, published his multivolume summa, *Historia Natural and General de las Indias*, between 1532 and 1546. Oviedo was a transitional figure in Western scholarship, with one eye on the ancients and his other on the present, and for both a wish to read originals. "For I do not depend upon the authority of some poet or historian when I write, but upon myself as an eyewitness of most of the things I shall speak of here; and what I have not seen myself I shall relate from the accounts of trustworthy persons, never depending upon the evidence of a single witness, but upon that of many, in those things I have not experienced in my own person." His final volume appeared two years after Copernicus published *De Revolutionibus* and 18 years before Galileo was born. This was natural history done by someone trained in the emerging genres of Renaissance humanism and someone who had begun his writing career by penning a chivalric romance. If it more resembles Theophrastus than Darwin, if it omits much that modern concerns and sensibilities wished it included, it was a monumental work without which our understanding of the era would be much impoverished.[6]

Not least, there was Bartolomé de Las Casas. Here was someone who had lived most of his adult life in New Spain, and who applied his mental energies to its description. His father had been with Columbus on his second voyage (1493); he himself arrived in 1502 while Hispaniola was still in upheaval, and from there he participated in the conquest of Cuba and laid plans to establish a model settlement in Puerto Rico. Later, revulsed by what he saw, he took holy orders, became a Dominican friar and eventually the first Bishop of Chiapas, and matured into the great collector of relevant texts on the human history of New Spain. In 1527 he began the first of three volumes of a *General History of the Indies*, finally completed in 1561. For his efforts he was widely regarded and reviled as either crank or crusader, and of course he despised Oviedo and other rivals, who returned his scorn.

Equally settler, scholar, and activist, Las Casas bequeathed a unique legacy. Sword and Cross, conquest and conversion had proceeded like two blades of a scissors, and together they had resulted in the near-total destruction not only of the native populations but of their religion and heritage, during which temples were torn down and codices burned. Amerindians lay outside the known pantheon of peoples: incredibly, Las Casas and others had to argue that they were truly human, that they had souls, that they deserved the protection of His Most Catholic King. If the civilization demanded that people could only be understood and absorbed legally through texts, then someone needed to create those texts. Las Casas was that someone.

The known peoples in the world were those recorded in the Bible, and in the classic geographies and histories of the ancients, particularly Herodotus and Strabo. To these were added other peoples adjacent to Europe, such as the Sami of the north, the nomads of the steppes, the semimythical folk of Cathay, the Indies, the empire of Mali, and Prester John. The best-known tribes were those nearest, all of whom were hostile and nearly all Muslim. The Iberians instinctively divided the world into Christians and Moors, sometimes with wildly inappropriate consequences. When in Calicut, da Gama passed a Hindu shrine and assumed, because Muslims did not tolerate images, that the statue must represent "Our Lady," the product of an unknown Christian sect. The newly discovered peoples stood outside the recognized concourse of humanity and the protocols evolved to temper relations between them and Europeans.[7]

As in so many matters, the transition was the Canaries. The Guanches spoke a new language, lived outside trade routes, were neither Christian nor Muslim. The classification of peoples mattered because it determined whether they were civilized or barbarian, whether natural law might apply to them, whether they might be overcome in a just war, whether they might be converted or not, whether they might be turned into slaves. They forced a discussion, already well developed in medieval times but from a feeble range of data, about what makes a creature human or civilized into a wider discourse. The discoveries of hundreds of previously unknown peoples in the New World especially posed issues in understanding as well as governance; and they could put church and state at odds. But as Felipe Fernández-Armesto has noted, "It is hard to detect any significant elements in the debate about them, at least in the first half of the sixteenth century, that had not been anticipated in discussions of the Canary Islanders." Yet the saga of their encounter seemed so mythical, the scale of discovery so much greater, the crisis of governance so much more pressing, the sheer shock of the Amerindian so sudden with its Otherness, and the means of disseminating the debate by printing press so effective that the old discourse was practically forgotten in the rush to the new.[8]

Some tribes disappeared almost as soon as they were discovered or pacified—the Guanches, the Arawaks, the Beothuks of Newfoundland (their ochre-daubed bodies the origin of *redman* as shorthand for America's indigenes). But the colonists also created new societies and peoples. Creoles complicated the politics of colonialism; mestizos, the organization of colonial society; and bands of frontier folk, the ability to impose a European-sanctioned order on tenuously held lands. Every line of contact had them: Cossacks in Russia; *degregados* in Portuguese Africa and Asia; *bandeirantes* in Brazil; *coureur de bois* in New France; bands of freebooters, pirates, and privateers everywhere. Typically a mixed-race group, they often proved useful as translators and cultural brokers. Almost as often, outside the grasp of law and state, they were provocateurs and a source of chronic instability.

Encountering affected both parties, and a degree of intermingling was inevitable. The mestizo society had its counterpart in mestizo systems of learning, even as a cultural creolism managed to retain its sense of superiority and universality while it absorbed words, concepts, places, goods, peoples, and practices from elsewhere.

At first the new discoveries fit into existing categories, enlarged and adapted. Then as novelty and numbers overwhelmed inherited classifications and cabinets of curiosities, they contributed to—compelled—an orbe novo of learning. The Great Voyages were not the only forcer. The edifice was already cracking from the Reformation, early-onset capitalism, and early modern states.

Scholars struggled to put the swollen caches of the new into the templates of the old. They appealed to the ancients, especially Pliny the Elder, whose *Natural History* made him the encyclopedist of Rome, and they amassed similar sprawling histories, both human and natural, and in some cases, galleries of images. In New Spain Francisco Hernández translated Pliny's 37 volumes while at the same time he relied on native guides and even used Nahuatl nomenclature, which he regarded as more accurate. But scholars dismissed his use of native terms; the Crown hid his collections and illustrations to prevent rivals from accessing them; and his Renaissance reply to Pliny was not published until the 1880s, and then as an antiquarian memento of Spain's imperial adventure. (Similar scenarios played out even for figures like Fernández de Oviedo, specially appointed to the task by the Crown.) At the time Hernández and Oviedo gathered their collections and interviews, modern science was still inchoate.[9]

The dates demonstrate how the world of modern science was out of sync with the world of geographic exploration. By the time Copernicus published *De Revolutionibus* (1543), De Soto and Coronado had completed their expeditions and the Portuguese had reached China and Japan. By the time Galileo was born (1564), Camões was eight years away from publishing *The Lusíads*, and the Portuguese Estado da Índia had passed its apogee. When Galileo published *Starry Messenger* (1610), a year after inventing the telescope, the Portuguese empire was perhaps 30 years beyond its prime, Spain's empire had passed its solstice, England and France had only begun tiny trading posts on mainland North America, though Holland had created the Dutch East India Company, the primary vehicle for interloping on Portuguese traffic, and Francis Drake had seized the Manila treasure galleon on his circumnavigation. Already, voyaging involved more piracy than exploration. Isaac Newton was born in 1642, the year Galileo died; his summa, *Principia Mathematica*, was published in 1687, and *Opticks* in 1704. By then only three new Pacific islands had been discovered over the previous 50 years. Australia and New Zealand were known only on their west coasts. The great satire on exploration and empire, *Gulliver's Travels*, was published eight years before Linnaeus's *Systema Naturae* introduced the modern systematics that would organize the natural history of discovery.

Modern science contributed almost nothing to the Great Voyages, save some navigational aids and cartographic conventions. Geographic discovery and scientific discovery were misaligned by a good century; scientists followed the conquistadores, mopping up the uneconomic crumbs. Still, trading factories set up research gardens and sought useful plants, and wealthy collectors gathered cabinets of curiosities. Hans Sloan's collections, largely from the Caribbean, became the nucleus for the British Museum; Kew Gardens evolved from the merger of two royal estates, which held formal gardens, some out of foreign flora. Empires of trade created a global matrix that forced collectors and intellectuals to reformulate the scale of the world and their understanding of how its many parts worked.[10]

Theophrastus, a student of Aristotle's, listed some 500 useful species of plants in his *Historia Plantarum*. By the time Linnaeus classified known plants for the many editions of *Systema Naturae* (1735–68), that number had increased by more than an order of magnitude; he apparently believed up to 10,000 existed on Earth. This was proportionately a major expansion of Europe's botanical data set, and it compelled new thinking about how to organize the largesse, both intellectually and institutionally. But the numbers named were a pittance compared to what remained to be found. Linnaeus's fabled summa of botany was hardly the equivalent of what the Great Voyages had achieved in general geographic discovery, and he based his organization on a concept from antiquity, the great chain of being. Fifty years later Alexander von Humboldt alone would ship back 60,000 specimens from South America, of which some 3,200 were new, and added another axis to make a map and group them by natural assemblages. A century later, Charles Darwin and Alfred Wallace, explorers both, added history.

While the First Age prodded modern science, not until exploration revived in the latter 18th century and metamorphosed into a Second Age, however, did they bond, one of many phase changes in the immense secularization that swept over Western civilization. Until then little changed. Two centuries after Portugal reached Japan, Linnaeus included only a few Japanese plants in his catalog. One of his apostles, Carl Peter Thunberg, eventually sailed as a physician on a Dutch East India Company vessel to Nagasaki, but he was denied the freedom to explore the countryside and had to rely on native collectors; not until 1784 did his *Flora Japonica* appear. By then Japan's Christian century was two centuries in the past. Not for another hundred years and after the Meiji revolution did the botanical exploration of Japan begin.

There was not much penetration of exploration into intellectual culture beyond law, theology, navigation, and cartography. There was negligible influence on art. In what may come as a surprise for moderns, science was neither a stimulus for the Great Voyages nor a primary benefactor from them. Science and geographic discovery developed along parallel tracks. They ultimately crossed, with vast consequences, but the advancement of knowledge in botany, zoology, geology, and anthropology—disciplines that did not even have their modern names much before the 18th century (and science, not until the 19th)—was neither a purpose nor an intended byproduct of the Great Voyages.

Where the larger culture and exploration did converge was in literature. Character, conflict, exotic settings, all valenced by narrative—this is the classic formula of the romance, which is often itself a graft onto the quest saga. There is plenty of evidence that chivalric romances influenced the temperament and perspective of Renaissance explorers, as technological romances have enthusiasts for space exploration in the 20th century. Unsurprisingly, writers adapted existing genres—the epistle, the epic, the essay, and of course the romance. Explorers were Jasons and Odysseuses; their story had Olympian dimensions; golden ages were relocated from the past to yet-to-be-discovered isles.

Travelers' tales crossed forms, sometimes hardening into factual guidebooks, often blurring into fantasy. As Prince Henry had carried the Crusades to the Western Sea, so had the knight errant, who pursued his quests in exotic lands newly discovered. Explorers viewed what they saw through the prism of their education and popular literature such as the *Amadís de Gaula*. Yet it was not always possible to sort one from the other—even the explorers could be confused. In an age of startling discoveries, it is no surprise to see the boundaries between romance and reality blur. Who could have invented the sacking of Tenotchitlan and Mallaca? Who could have conjured up the travels of Francisco de Orellana and Fernão Mendes Pinto? Who could have forseen the voyage of the *Victoria*?[11]

The project acquired a technological boost from the invention of movable print, as critical for the dissemination of knowledge as the caravel for voyaging. Johannes Gutenberg set up his first press in 1450, and printed his iconic Bible in 1455, roughly the same year Portuguese marinheiros discovered the Cape Verde Islands. The Great Voyages and the press coevolved. Travel accounts were always popular, and the press became a way of popularizing discoveries, at least those not locked in state vaults. Many of the early explorers released preliminary reports in the forms of letters, which were quickly copied, printed, and distributed. Books flooded Europe as New World bullion did Spain. In the 11th century Europe had perhaps 200,000 manuscripts, both originals and copies; in the 15th century it had an estimated 8,000,000 manuscripts and 11,000,000 books. In the 18th century, as the First Age settled into lassitude, it had a billion.[12]

A special category of writing belongs to chroniclers, aggregators, and propagandists. They fall rudely into two groups, those who recorded exploration and those who sought to promote it. The first is deftly distilled into the career of Peter Martyr d'Anghiera, chronicler for Spain's Council of the Indies and author of the *Decades*. Countries more on the margin of exploration inclined toward cheerleading, not just chronicling. This was the case, most famously, with England. Its first press arrived with William Caxton in 1477 and specialized in what the public wanted to read; works of geographical discovery were not among them. Despite Bristol merchants, England's interests pointed to Europe rather than New Found Lands, which were abandoned after the Cabots' voyages, save for fishing. Not until 1554, with the crisis of Mary Tudor's marriage to Philip of Spain, did the *Decades* appear in English. It did so amid a political crisis, economic opportunities, and a spiritual call to arms. Protestant England now joined the competition against Catholic Spain.[13]

The critical voice was Richard Eden, England's "first literary imperialist." Eden began by publishing *A Treatyse of the Newe India* (1553), arranged to have the *Decades* translated along with Martin Cortes's *The Arte of Navigation*, urged his countrymen to learn from the major colonial powers, and then expand for themselves, searching out new markets for the "commoditie of our countrie," which was wool. In his reasons for travel he restated God, gold, and glory into serving God, improving the national economy, and inspiring an entrepreneurial and adventuring spirit essential to a robust people. If England wanted to be more to Europe than Ireland was to England, then it had to expand its horizons. The best defense was a forward strategy of aggressive exploration. The earliest endeavors focused mostly on the Northeast Passage, culminating in the Muscovy Company and trade with Russia, and then with renewed interest in a northwest passage. But as Eden's first book suggested, the Indies were the prize, and as the public reaction noted, the governing classes and reading public were more inclined to look inward than outward.[14]

That changed in 1580 when Spain absorbed Portugal, reducing part of the dynamic rivalry that had powered the voyages and helping set the conditions for others. England concluded two routes to the wealth of the East: a trade treaty with Turkey and the astounding return of Francis Drake from his "world encompassed" circumnavigation, not least with the *Golden Hind* stuffed full of the plunder from unsuspecting Spanish ships and towns in the Pacific. Humphrey Gilbert experimented with an American colony. And England found its great publicist of voyaging as Richard Hakluyt, then thirty years old, commenced his pleas for an English overseas empire of commerce and colonies.[15]

An older Hakluyt recalled that, as a youth, he had chanced upon "certain books of Cosmography, with an universal Map" at his cousin's study, and that he then and there resolved that he would "by God's assistance prosecute that knowledge and kind of literature." His 1580 appeal was followed by a 1582 compendium, *Divers Voyages Touching the Discovery of America*, in which he exploited historical events to argue the cause for English expansionism. The *Particular Discourse on the Western Planting* followed, this time aligned with schemes by Walter Raleigh. More than a simple shill, Hakluyt was a Renaissance scholar, versed in many languages and intent on amassing accurate accounts from any source and thus eager to talk with pilots, merchants, and captains. His masterpiece followed the year after Elizabethan England defeated the Spanish Armada and Thomas Cavendish returned, like Drake, from a lucrative circumnavigation based on piracy in Spanish America.[16]

Principal Navigations, Voyages and Discoveries of the English Nation was an immense encyclopedia of travel that celebrated the new age of adventuring while seeking to establish continuities with an ancient English past. It argued that England was, by heritage, a voyaging and trading people, that an expansion of enterprises was both possible and essential, and that accurate knowledge was the font of such inspiration. The *Principal Navigations* was in equal measure patriotic and practical. A second, expanded edition published in 1598–1600 weighed in at a million and a half words distributed into three volumes. Here was history in the service of commerce and politics. For Elizabethan England Hakluyt was to exploration what Shakespeare was to theater.

The outcome could not help but awe, and through awe, to inspire both an admiration over what has been achieved and an ardor to further it. Yet Hakluyt was candid about his purpose: "our chief desire is to find out ample vent for our woollen cloth, the natural commodity of this our realm, the fittest place I find for that purpose are the manifold islands of Japan and the northern parts of China, and the regions of the Tartars next adjoining." To this goal were gradually added colonies in America, where trade might be supplemented by the natural wealth of those lands. The second edition made the point even more explicit by adding "Traffiques" to "Navigations" and "Discoveries."[17]

Others like Samuel Purchas (*Purchas His Pilgrims*) were less fastidious about authenticating sources and less willing to subject readers to vast catalogs of prospective trade goods. They abridged the long chronicle into a brisker narrative, pumped up the emotional receipts relative to empirical expenditures, and bequeathed a saga that made the voyage of discovery and the planting of colonies a distinctive and necessary feature of the English identity. Complicated accounts of commercial traffic became tales of derring-do and patriotic glory. The exorbitant costs were only apparent: the enterprise, readers were assured, would pay for itself many times over.

For some journeys, mirages work as well as landmarks. Samuel Eliot Morison observed that "although the French seaports had every possible advantage for Atlantic exploration that the English had; although the French crown gave its merchant marine more support and encouragement than the Tudors ever did, there is one English asset which the French lacked—a Hakluyt." They lacked that enormous compendium, the comprehensive vision, that sense of urgent inspiration. Later, enthusiasts for a Greater France lamented the similar lack of robinsonades—French versions of Robinson Crusoe; popular stories of overseas adventuring. Exploration, it seemed, was not built into humanity's genome. It had to be incubated, cultivated, and roused to action.[18]

In practice the scope for exploring was restricted. The Renaissance's reconnaissance had found what Europe believed it wanted. For England looting was more lucrative than trading wool, and for England, France, and Holland picking off Spanish and Portuguese claims was more productive than raw exploration into still-unknown territories and uncertain trade relations.

Contrary to Western vanities, the discovery of Europe might mean little, much less than Europe's discoveries meant to Europe. They would have to fit into existing cosmologies and institutional arrangements. Europe was a small piece of a large continent, a minor power amid sprawling empires, a backwater in global geopolitics. Its knowledge of nature and Earth made it a lesser civilization; its perennially quarreling principalities made it at best a regional power; its economy had little to entice the world's bustling entrepots. They were less interested in Europe than Europe in them. What Europe boasted about most was its possession of the True Faith and its technologies, of which military hardware was foremost, and later of its sciences. The first few others wanted, so long as their old creed thrived; the second, they learned to respect and ultimately crave.

The West on the eve of discovery wanted to know. It was ravenous for knowledge, not as deeply or as widely as it was greedy for gold, but the differences between discoverers and discovered involved more than gunpowder; and in time the desire to understand, to answer geographic puzzles, linguistic riddles, and ethnographic curiosities became a motive for extraordinary undertakings. That took time. The First Age had a ravenous curiosity; the Second learned to discipline it.

Until European discovery led to European hegemony, there was, even among its keenest rivals, not much concern about Europe. Europe cared more about the wider world than that wider world about Europe. Since late

medieval times Europe had hungered after learning, initially by translation from ancient Latin and Greek texts, then from Arabic texts that were
themselves translations from antiquity. The process of geographic discovery
quickened curiosity, not least about languages, and this among missionaries
as well as secularists. The latter 18th century stands outside the First Age,
but it enjoyed the fruits of that earlier efflorescence. Bernard Lewis notes
that by then "some seventy books on Arabic grammar had been printed in
Europe, about ten for Persian, about fifteen for Turkish. Of dictionaries there
were ten for Arabic, four for Persian, and seven for Turkish." Many were not
themselves mere translations but original scholarship. "There was nothing
comparable on the other side." The "first bilingual dictionary of Arabic and
a European language by a native Arabic speaker was published in 1828. It
was the work of a Christian—an Egyptian Copt—'revised and augmented'
by a French orientalist and, according to the author's preface, was designed
for the use of Westerners rather than that of Arabs." A similar project was
underway in India, where European scholars worked out the genealogy of
Indo-European languages.[19]

The cultural story, in brief, is not just one by which Europe imposed its
values and economies on others. In the case of the East, Europeans were the
humbler civilization that knew less, possessed less, and wanted more than
those they encountered. Yet just as it was Europe that set out on the Great
Voyages, so it was Europe—seemingly as keen for knowledge as for spices
and souls—that sought to learn disproportionately about the wider world.
The unstable amalgam of ambition and desperation that launched ships also
set Western learning on its own grand voyage of discovery. What it learned
then affected what it found.

How then did knowledge and exploration relate? There are some cases of
direct transfer, notably regarding geography, in which discovery enters into
the formal literature of scholarship and statecraft. There are instances of
exploration as a symbol of a new world a-borning in Europe and for learning. But mostly they were part of a shared culture undergoing a profound
chrysalis.

When rocks are subjected to heat and pressure, they metamorphose. Shale may reconstitute into slate, and then into schist. The ingredients—the essential chemicals—remain the same; they just assume new forms, and in extreme case may melt down into gneisses and granites. Something like that happened during the era of the Great Voyages. Politics, commerce, literature, painting, theology, and almost all the rest of Europe's cultural ingredients underwent the heat and pressure of desire, challenge, threat, and rivalry and recrystallized that basic stuff into new forms and arrangements. Some parts were more fully assimilated than others, and some less, like blocks of country rock suspended in schist, yet they all shared a common experience and so resembled one another as well as what they descended from.

The Reconnaissance was never isolated from all the rest of what was occurring throughout the extended Renaissance. But what had been pilgrimage, travel, war, trade, and sheer wandering had partly melted and reformed into something recognizably new and part of a larger reconstitution of European civilization. Along with commerce, statecraft, new technologies, and a bottomless hunger for something more, the wanderlust that recrystallized into exploration aligned with other features to propel the rise of the West.

10

The Great Conjunction

What is greatest and most memorable of all, you have brought together under your command peoples whom nature divides, and with your commerce you have joined two different worlds.

—PIETRO PASQUALIGO, VENETIAN DELEGATE TO KING MANUEL OF PORTUGAL[1]

Empires had previously come and gone, often over the same lands, as though human history obeyed analogous rhythms as those that governed natural history. But this new wave of expansion was different: it joined what had been sundered across oceans and over geologic time, and the effects could not be reversed. It involved not just new rulers and rewritten maps but a remaking of the planet. The evolving arrangements for world-spanning trade and capital rewired global cycles of chemicals and organisms, broke and fused biotas, found and rerouted energy pathways, and reconstituted human demographics. These reforms could not be replaced or recycled in the way of previous empires. The emerging infrastructure for global trade and capital remade the infrastructure for the global biosphere. What Europe did by connecting maritime cultures it also did by joining hearths for domestic flora and fauna. The great conjunction of planets in Prince Henry's horoscope was manifest as a biotic collision between new and old worlds.

For good and ill, European exploration—what Alfred Crosby called the Columbian exchange, and later ecological imperialism, and Richard Grove green imperialism, and others might consider a prologue to the Anthropocene—not only opened Pandora's box but carried it everywhere save the ice sheets of Antarctica.

As with so much of the First Age, part of the project was deliberate and part, probably the largest, the outcome of accident, coincidence, and opportunity. As Crosby notes, the newcomers did not travel alone: they carried with them a mobile, "portmanteau biota" of servant species, retainers, camp followers, and stowaways—ecological allies—that catalyzed the transformation of new lands into something like the older lands they came from. Ecology also had its encounters.[2]

The deliberate dimension involved finding and relocating useful plants and animals. Some were familiar but valuable staples like sugar and cotton which could be planted in suitable lands unveiled by discovery; others were novel, like potatoes and chilies that could be brought back from newly found lands and planted in Europe. Medicinal plants were always sought. Trading factories often had gardens to collect and experiment with local varieties; imperial ambitions elevated them to royal institutions like Kew Garden and Jardin des Plantes. Both impulses—to find places to grow species already known and to find valued novel species—were objectives of exploration. The project was, for naturalists, the biotic counterpart to humanists' quest for finding and translating texts.

The ultimate quest was for gold. But trade species like cinnamon and pepper could be as valuable as gems. The Great Voyages thus coincided with what John Richard has called the Great Hunt. Europeans sought out furs, whales, seals, fish, pastures, timber, hides, and meat as well as bullion. Grass and gold sent prospectors around Australia; martin and silver fox drove Russian promyshlenniki across Siberia and to Alaska. Birds as specialized as the dodo disappeared; ungulates like bison that numbered in the tens of millions could teeter on the brink of extinction. In return cattle, sheep, and house flies abounded. And then there were those organisms that were the fellow travelers, stowaways and opportunists, like Norway rats, *Anopheles* mosquitoes, ragweed, and smallpox.[3]

Much as Europeans took over the carrier traffic in the seas they reached, so they rerouted the passage of ecological goods, such that bananas and rice went from Southeast Asia to Africa and America, and cassava could voyage from the Caribbean to sub-Saharan Africa without passing through Europe, and the pineapple might go from Brazil to Hawaiʻi. Madeira and the Canaries were, once again, proving grounds for colonization based on agricultural goods, notably sugar. Large swathes of temperate and Mediterranean-climate lands became, in Crosby's phrasing, Neo-Europes. Apart from their contribution to knowledge, apart from their role in the rise and fall of empires, the Great Voyages expanded the realm of humans, even if much that transpired remained within reach but beyond grasp.

Mostly the Americas exported plants and imported animals. There were few domesticated animals—dogs, turkeys, guinea pigs, and camelids like llamas and alpacas. But there was a cornucopia of domesticated plants. Animals spilled out of Europe like a Noah's ark: cattle, horses, pigs, sheep, chickens, oxen, cats, burros, and multiple breeds of dog. The plants that made the translation were mostly those Europe had acquired from old trade routes to the Middle East like cotton, sugar, citrus, grapes, and such cereals as rice, wheat, oats, and barley; but there were also fruits like apple and apricot and the pasture grasses the émigré livestock preferred and that came with them, and there was no end of weeds. In return Europe got tomatoes, potatoes, chilies, squash, pineapples, mangos, papayas, chocolate, avocadoes, and maize, which it exported around the world to its settlements and trading outposts. The First Age didn't simply replace rulers, or human inhabitants. It reconstituted habitability itself.[4]

Today, it is almost impossible to imagine what cuisines were like before the Great Voyages: Northern Europe and the British Isles without potatoes; Italy without tomatoes; Europe and its Neo-Europes without coffee, tea, and cocoa or the sugar to sweeten them; Mexico without wheat, pork, chicken, beef, mutton, lettuce, citrus, and new varieties of beans; Africa without cassava, maize, bananas, and rice.

The most obvious reconstruction of the world was the biogeography of its humans. In the Americas, in inhabited islands, in Australia, the indigenous populations collapsed. The lands were occupied by emigres from Europe or the Neo-Europes, by slaves, convicts, or other coerced colonizers, and by mestizo societies. Today Europeans and Africans are in places that, by traditional migration, they would not be. They are the human version of the ecological exchange that had remade the biosphere.

The reasons are many. The core, however, seems to be the role of diseases and their vectors as part of the ship's complement that did the voyaging. Simply by their presence Europeans introduced measles, smallpox, tuberculosis, and flu, and by their behavior cholera and plague. In the Americas the exchange went almost only one way: from the Old World to the New. Europeans had themselves endured waves of plague over the centuries, and had places from lowland England to Sardinia, that had endemic malaria, and a portion of its surviving population had acquired some immunity. The Amerindians had none. There is a good argument that such disease frontiers helped small bands of conquistadores to overcome what had, on first contact, been teeming populations of indigenes. Again, the Canaries may have provided the precedent, as *la peste* broke the capacity of the Guanches to continue to resist as they had for most of the 15th century. The conquest of Mexico may well have been preconditioned by smallpox introduced by the earliest Spaniards. Certainly, North American natives tended to melt away in advance of settlement.[5]

But the reverse could hold elsewhere, and nowhere more than Africa. Here diseases tended to go the other way, although it may be that smallpox and some respiratory diseases helped open central Africa to Swahili and European slavers. Otherwise Africa's dense disease environment made penetration almost impossible, not only for Europeans but for their livestock. "Fever" of one etiology or another was inevitable: Europeans could colonize only in Mediterranean or temperate climates or select inland plateaus. Where Africa contributed to a global diaspora was through the slave trade, which transported diseases such as malaria and yellow fever (along with the appropriate mosquitoes) to the Americas. The relative health of black Africans encouraged more slaves; the turnover of European garrisons in the Caribbean was nearly 100 percent per year.

Where Europeans could thrive, they tended to rule by demographic takeover as well as by sword. Where immigration could never hope to rival indigenous population or where diseases pummeled the migrants, they ruled indirectly, assuming the institutions of the previous ruling class and working through local chiefs.

Terraforming was not unique to Renaissance Europeans. People moving into new territory have always remade the encountered lands into something more suitable to them. That is how agriculture came to Neolithic Europe, how Bantu peoples slashed, burned, and herded down Africa, and how Polynesian colonizers, their boats laden with taro, pigs, and rats, had ecologically transformed Pacific islands, and wiped out flightless birds on New Zealand well before the Dutch did in the dodo on Mauritius. What Europe contributed was to boost the process into a planetary scale, further accelerated and broadened with the later advent of fossil-fuel powered industrialization, for which the European imperium served as vector.

Yet Europe also contributed a self-critique of this transfiguration. One perspective celebrated the new settlements as progress, as a way of making "waste" lands fertile and more productive. Such places were converted by the plow and axe from barbarism to civilization. The other view framed the narrative of change according to the Biblical account of the Garden and Fall. The new lands, particularly isolated islands, were an Eden that Europeans through bad behavior, ignorance, and sheer blundering turned to a desert. As with so much of the First Age, this was an old trope, inherited from antiquity, now translated and glossed to accommodate the realities of a geography well beyond the eastern Mediterranean familiar to the ancient philosophers and the Levant known to Old Testament prophets.

In the era of the Great Voyages the debate devolved with special force on Hispaniola. Columbus expounded the Edenic version upon returning from his first voyage. The land was "very fertile to a limitless degree," the harbors "beyond comparison," its mountains greater than Tenerife, its trees "of a thousand kinds and tall, and they seem to touch the sky"—palms, pines, fruits. There were birds abundant, including singing nightingales. The interior was filled with precious ore. The native population was "without number" and prelapsarian—naked, timorous, and "so guileless and generous with all they possess, that no one would believe it who has not seen it." The island could produce spice, cotton, aloe wood, rhubarb, cinnamon, other spices, "as much gold" and "as many" slaves as Spain might wish. (There was a grimmer presence in the form of "very fierce" cannibals that ranged through the islands, but they were not the resident population.) Hispaniola, he concluded, was "a marvel." It was "a land to be desired and, seen, it is never to be left."[6]

Such Edenic sentiments were applied to a few newfound lands—to Atlantic isles, the Americas, and later, Pacific isles. There was no African or Asian, or Australian equivalent. Africa and Asia had lands that were often more densely and more complexly settled than those of Europe, were not "new" in the same sense that Hispaniola and Madeira were, and could not fit tidily into a Biblical matrix. Australia appeared, as Charles Darwin once put it, as though it were a separate Creation, and certainly to European eyes no Garden. The Edenic vision pointed to an earlier, more natural time; not until the lead-in to space travel did the Garden seem to lie in a future, more technologically sophisticated civilization.

Columbus's first letter was, in its way, the environmental equivalent of a chivalric romance. Here was an idealized account that hedged into the delusional. Columbus was unable to found successful colonies, or even site his remarkable discoveries in the real world, but he did succeed in penning a durable genre for describing a new world. Those fruited trees, however, had their serpent, and it did not take long until the Fall. The native Taino population collapsed, and without their tending, the Garden went to weed. (African slaves were eventually imported to replace the vanishing native laborers.) The European population was unruly. A place of bustle and usufruction became a breeding ground for lethargy and a holding pen for ill-tempered conquistadores.

The counternarrative came from Bartolomé de Las Casas. His celebrated polemic, *A Short Account of the Destruction of the Indies*, written in 1542 and published in 1552, portrayed the conquest as thuggery, transforming what had once been a beehive of productive habitation into a blasted shell. The ruination of the indigenes had not only deprived New Spain of labor it desperately needed (and souls it said it wanted to save) but trashed the landscape, leaving heaps of biotic slag. A garden went to weed. An Eden became a desert. The New World wilderness was not something Columbus discovered but something his blundering helped create.[7]

As with so much of exploration, themes persisted, even as they transmigrated. The advent of capitalism commoditized nature. The maturation of science desacralized it. Advocates like Galileo urged scholastics to turn from their musty libraries to the book of nature. Unlike most chroniclers, the Book of Nature recorded eloquently and starkly the changes wrought by contact, and both within its lines as well as in the glosses between them it was willing to tabulate the losses as well as the gains. Whatever explorers thought they had found, their actions typically transformed into something different.

The Discoverers and the Discovered

And there I found very many islands filled with people innumerable, and of them all I have taken possession for their highnesses, by proclamation made and with the royal standard unfurled, and no opposition was offered me.

—CHRISTOPHER COLUMBUS, *LETTER ON FIRST VOYAGE*[1]

D iscovery unsettled more than intellectual systems. Exploration could segue into trade, and even empire; it could upset the balance of power within an exploring nation and between it and European rivals; it demanded that encounters with indigenes evolve into formal relationships. It had, that is, political, legal, and moral facets that required some kind of institutional regime to prevent anarchy. Fundamentally, it required a system, codified and recognized by all parties as legitimate, to define both the rights of the discovers and those of the discovered. It needed to keep the discoverers from fighting among themselves and to guide them in their dealings with those they discovered.

The Romans had prescriptions for dealing with the many peoples they met and ruled, codified into a *jus gentium*. As with other matters, Renaissance jurists began with that recovered notion, but as so often, they found antiquity's wisdom rapidly overwhelmed by events, which soon outstripped understanding and institutions. The Reconnaissance happened very quickly and across a globe at a time when simple letters could take weeks to cross Europe, when the modern state was still inchoate, when traditional relationships were unmoored by the discovery of whole new worlds of peoples. A legal framework came decades, and in some instances, centuries after new facts had been created on the ground. Even before the Great Voyages, however, assorted norms and rituals were invented and tested in the Atlantic isles and along the African coast.

Prescriptions evolved on what first discovery meant. Sometimes it was enough to sight a new island, but with so many mirages floating in the sea—so many sightings that were never confirmed by anyone else—the usual ritual was to make a landing, read a proclamation, and leave some material artifact behind, often a stone cross (such as the Portuguese padrãoes that lined the coast of Africa). Publishing the event announced the discoverer's claim and warned off others. If an explorer did not claim outright sovereignty, he could claim exclusive rights to trade. Since exploration was by ship, a claim extended to the surrounding sea as well as over land.

The process didn't stop Spain and Portugal from quarreling over islands and coastlines in Guinea. When the Treaty of Alcáçovas confirmed Portuguese rights to the West African trade, and as John II ascended to the throne, the voyages of discovery were renewed. That quasi-domesticated squabbling came unhinged with Columbus's discovery of islands far to the west. Negotiations soon began, with an appeal to the pope as arbitrator, which in 1494 led to the Treaty of Tordesillas. The treaty divided the newly unveiled world along a line of longitude 370 leagues west of the Cape Verde Islands. Portugal had claim to everything to the east, Spain, everything to the west. This confirmed what Portugal most wanted—the route to the Indies that it was certain it would find in the coming years. Spain would get the new lands being surveyed by Columbus, which Portugal was convinced were not the Indies. Six years later Pedro Cabral landed in what he named Brasil, which lay east of the line, and so became Portuguese. Further discoveries elaborated on the extent of the New World, which Spain dutifully claimed along with the adjacent seas. While the treaty prevented an Iberian rivalry from sprawling across the globe, no other European nations honored it, though none were in a position to challenge the first movers either.

Then Portuguese explorers pushed beyond India to the Moluccas themselves, and soon afterward Spain sent the Armada de Molucca across the

Pacific (and back). In which half of the world did the Spice Islands lie? No one knew, or given the state of cartography and navigation could know. A panel of experts was assembled, and agreed on how the dividing line should look on the other side of the globe. The result put the line almost 300 leagues to the east of the Moluccas. This was confirmed with the Treaty of Zaragoza in 1529. Portugal got what it wanted—exclusive claim to the core Spice Isles, China, and Japan. Spain was unwilling and unable to contest, though it later, on the basis of Magellan's discovery, colonized the Philippines, which lay within Portugal's sphere. The project helped contain the rivalry. Again, other nations ignored the treaty, and happily picked off whatever posts and trade they could. And of course the discovered peoples, many of whom were vastly more numerous than the discovering nations, had no say in the matter.

So the newcomers were pleased to ignore the Iberian legal regime, especially if they were Protestants and could conflate their commercial and religious competitions. A century later they felt the need for a more comprehensive system. The Dutch led, since they were the most aggressive partisans. Their reconsideration produced two major works, both by Hugo Grotius, one of the great Renaissance jurists and not incidentally an advocate for positions that favored the new order over the old.

Mare Liberum (*Freedom of the Seas*), was published in 1609. It argued that, away from coasts, no one could claim the ocean. The high seas should be free and open to all. It had its own laws, with ships subject to whatever national law they sailed under. Here was a doctrine to disarm conflicts over who might control trade on the open ocean, and it favored a state, like the Netherlands, committed to maritime commerce. Where seas and lands (or islands) intermingled, as in the Mediterranean, and one could sail across the widest sea in a few days, it made sense that land and its adjacent sea could be treated as a single element. Where the Ocean Sea spilled across the globe and months might be required to sail between ports, it did not. *Mare Liberum* segregated the law of the sea from the law of land.

Grotius published his other master work, *De Jure Belli ac Pacis* (*On the Law of War and Peace*), in 1625. It revisited the notion of a just war, the right of conquest, and the legitimate fruits of conquest. An unholy or unjust war could invalidate the right to rule. Clearly, this was another gloss on a world no longer controlled by an Iberian duopoly, particularly once Portugal had been absorbed into Spain. What had been justified on religious grounds now had some foundation in secular law, however much such notions lagged behind the times and were ignored. The reality remained what William Wordsworth put into verse years later as the "good old rule . . . / That they should take, who have the power / And they should keep who can."[2]

Still, if incrementally, exploration was forcing an intellectual reorganization to match the shifting arrangement of power on the ground. It nudged a reconstruction of the jus gentium into what became international law. Exploration might spill out from the confines of Europe, but that did not mean war had to follow it.

Redefining the relations between discovering peoples and those they discovered was more vexing. The first distinction was between lands that held people and those that did not. Especially where a place was already settled, the newcomers did not always seize the land; most often what they wanted was trade, preferably a trade monopoly. They were led into fights and land claims to support that trade. Still, they sought some legal grounding, or invalidated claims might lead to fights among themselves or with their European rivals. If they claimed by theft, they had no basis to protest its subsequent theft by others.

Uninhabited lands were easy. They were a *terra nullius*, a land of no one. It was a case of finders keepers; first in use, first in right; it's no one else's, so it's mine. Inhabited lands were harder. Where Europeans met advanced civilizations, they had to negotiate by treaty or seize by conquest. Where they encountered people less technologically advanced, or whose occupation of the land was not founded on notions of fixed land ownership or on agriculture based on the plow, or met aboriginal peoples who hunted wild beasts rather than raising greater numbers of livestock, justification was trickier. The concept of a terra nullius was expanded beyond land not occupied at all to lands not occupied in ways that maximized their use. In other words, lands through which peoples moved seasonally, exploiting different plants and animals at different times of the year, or who had no notion of land ownership on the European model, also constituted a terra nullius. For legal and political purposes such places could be treated as uninhabited.

The thornier difficulties occurred when contact meant conquest, when the newcomers had to work with the indigenes as a vital labor source, when the indigenes were at first far more numerous than the newcomers. This was the dominant situation throughout Spanish America. How to justify an often brutal conquest? How to establish appropriate relations between conquerors and conquered? The colonies needed labor; the natives were the obvious source. The injunction to convert to the true faith could fold uneasily into forced conversion to labor. Events happened quickly, so it was not surprising that those on the ground should adapt the practices that had accompanied the Reconquista in Spain, and dole out natives in a neofeudalist manner through encomiendas and repartimientos. Indigenes were treated as serfs in fiefdoms, or sold as slaves, or as their ranks faded were replaced outright with imported slaves from Africa.

If Spain was the most dramatic example, it was also often its own best critic. If it needed to exploit the natives for labor, it was also charged with oversight for their souls. The pope had blessed the conquest so far as it brought souls to the Faith, preferably peacefully, but that charge placed the project in the realm of morals, and it allowed the church to question the undertaking in ways not possible for other institutions. Besides, the indigenes were also pawns in the power shuffles between king and nobles.

The questions inspired scholars, some uneasy or outraged by what was happening, and some, more grounded in abstractions, apologetic. Among the critics the outstanding academic voice belonged to Francisco de Vitoria, who, arguing from natural law, expanded the old jus gentium into a universal doctrine, what became a basis for international law (and in much later times, of human rights). The apologists found their voice in Juan Ginés de Sepúlveda, who (as J. H. Parry observes) trotted out "almost all the arguments ever cited in favour of imperialism." If Vitoria based his understanding of natural law on a shared humanity, Sepúlveda spun it into the case for an inevitable hierarchy, including a condition of natural servitude. Further, he declared that it was a right, perhaps an imperative, even by force of arms, to subject people "who require, by their own nature and in their own interests, to be placed under the authority of civilized and virtuous princes and nations, so that they may learn from the might, wisdom and law of their conquerors to practice better morals, worthier customs and a more civilized way of life." In 1542 he consolidated his views into *Democrates Alter*.[3]

What sparked an academic discourse into a political firestorm was Bartolomé de Las Casas, who wrote a bitter counterargument to Sepúlveda with one of the most famous polemics on colonization, *A Short Account of the Destruction of the Indies*, which denounced not just individual conquistadores but the whole project and ultimately questioned its legitimacy. Others piled in. Even those who agreed with Las Casas could find distasteful his full-frontal denunciations, which helped rivals frame a Black Legend of Spain. They doubted that substituting missionaries for conquistadores was more than utopian rhetoric. Someone (perhaps Sepúlveda) even denounced Las Casas to the Inquisition. More avid, flinging Las Casas's outrage back onto itself, were those like Bernardo de Vargas Machuca.

Vargas Machuca was part of a second generation to come from Spain, this time when Spanish America had to defend itself against foreign privateers as well as acquire more gold and expand the Faith. He engaged in various campaigns, held assorted posts in the colonies, and sought favor at the court, eventually becoming castellan of Portobello, Panama. His experience had left him "bankrupt, bitter, and in broken health." Between 1602 and 1618 he wrote a sharp reply to Las Casas titled *Defense of the Western Conquests*. His was a soldier's defense, proud and acerbic, the voice of those who had done the hard, dirty work of empire building. People who resisted the Faith deserved to be forcibly converted, people who resisted the conquest asked for retaliation, people who questioned the magnificence of the conquest deserved scorn. The empire had been won by the sword and had to be held by the sword.[4]

Machuca's *Defense* came 50 years too late to influence the overall debate. The Las Casas faction had already redirected the official trajectory of Spanish policy when Charles V promulgated the New Laws for governing the American colonies, which abolished the encomienda system and enshrined, at least in principle, other reforms. Both sides wanted more and each denounced the other as traitorous. In 1550 Charles V arranged for a formal debate in Valladolid between Sepúlveda and Las Casas. It was technically inconclusive, for although Sepúlveda may have won on logic and Aristotelian sources,

Las Casas scored on moral grounds. He returned to Chiapas as bishop and Protector of the Indians, and later retired to Spain to assist the Council of the Indies with policy.

Conflicted Spain reflected the larger tensions within Europe. Over time, to many of Europe's intellectuals the indigenes could appear more innocent and virtuous than their nominal superiors. Over the centuries exploration could seem a wondrous adventure; but empire, its dark double, could act as an unflattering mirror held before European civilization. The Black Legend could drip its stain onto other nations as they tried to seize empires for themselves. Two centuries later in *Gulliver's Travels* Jonathan Swift reduced the ritual of discovery to savage satire:

> To say the truth, I had conceived a few scruples with relation to the distributive justice of Princes upon those occasions. For instance, a crew of pirates are driven by a storm they know not whither; at length a boy discovers land from the top-mast; they go on shore to rob and plunder; they see a harmless people, are entertained with kindness; they give the country a new name; they take formal possession of it for their King; they set up a rotten plank or a stone for memorial; they murder two or three dozen of the natives, bring away a couple by force for a sample, return home and get their pardon. Here commences a new dominion acquired with a title by Divine right. Ships are sent with the first opportunity, the natives driven out or destroyed, their Princes tortured to discover their gold, a free license given to all acts of inhumanity and lust, the earth reeking with the blood of its inhabitants, and this execrable crew of butchers, employed in so pious an expedition, is a modern colony sent to convert and civilize an idolatrous and barbarous people.
>
> But this description, I confess, doth by no means affect the British nation.[5]

Of course not.

Like Sepúlveda's arguments for imperialism, both advocates and critics can sound uneasily familiar to contemporary ears.

Academic discussions, even when attended by a king, could not halt further exploration, which outstripped both ideas and institutions. The First Age had an internal dynamic that would take decades before pausing, and several centuries before it was exhausted. But they forced European civilization to judge what had happened and try to ameliorate the worst excesses. Mostly the discourse occurred after the sprawl of discovery; that it might have halted the Great Voyages altogether or before they happened is quixotic.

Even a national epic could not undo what had been done. Writing after the great surge that created the Empire of the Indies, channeling his rebuke through the Old Man of Belém, Luís Vaz de Camões knew that the wind in the age's sails was too powerful; that, although Portugal's great navigators will rival the exploits of Odysseus and Aeneas, they will also cause grief and pain. Yet the worry comes too late. The ships are already unmoored and riding the ebbing Tagus out to sea, the Old Man's voice is too feeble over the widening waters, his warning goes unheard and unheeded. The tidal pull of the times was too great.[6]

12

Ebb Tide

And though he said he would not cease to pray for me, yet he would venture to say to me, that if I did take this foolish step, God would not bless me, and I would have leisure hereafter to reflect upon having neglected his counsel when there might be none to assist in my recovery.

I observed in this last part of his discourse, which was truly prophetic.

—ROBINSON CRUSOE, RECALLING HIS FATHER'S ADVICE (1719)

The legacy of the Great Voyages began to crumble both from internal strains and external stresses. With their empires badly overextended, Portugal and Spain squandered the windfall of conquest with ill-conceived wars, and endured attacks from northern European competitors that evolved into a feeding frenzy. Neither saw the need for further exploration: they sought to hold what they had. Nor did their new rivals seek new lands: it was easier to seize existing ones from the Iberians.

The Portuguese lost many posts in Asia and the Atlantic. The Dutch East India Company founded a rival to Malacca at Batavia and took over the trading operations at Nagasaki. English and French counterparts created trading factories along the coast of Guinea and India. The Dutch erected a provisioning outpost at Cape Town along the carreira da Índia; for 20 years they seized and held Brazil. The constellation of posts that made the Estado da Índia drew European competitors, and slowly contracted, withering until only a few sites remained, like gems left after rock erodes. Poachers and pirates roamed the Spanish Main, raiding settlements and establishing their own factories beyond the reach of Spanish authorities—for France, Quebec (1610) and Guiana (1624); for England, Jamestown (1605) and Plymouth (1620); for Holland, New Amsterdam (1624). Aspiring colonizers began to bargain over colonies as chips in Europe's wars, which led to France getting the West Indies islands in 1625–50, and England, Jamaica in 1655 and other West Indies isles from 1627 onward. The claimants then fought among themselves. The Dutch seized Surinam from England in 1667; the English, New Amsterdam from Holland in 1664. All intervened in West Africa. Holland, Britain, and France quarreled over the North American fur trade.

Most of this was a transfer from one nation to another, not a renewed wave of exploration. There were a handful of fresh discoveries and factories along the margins, as rivals attempted to outflank each other. Mostly, one nation substituted for another the way petrified wood replaces lignin with chert. Enthusiasm for raw reconnaissance faded: rivals preferred to swap posts rather than quest for new Indias and Mexicos. Europe calmed its worst outbursts. Secularization seeped through the culture; modern science developed alongside religion, as though an institutional version of Descartes's separation of mind and matter. Wars of trade replaced wars of religion. John Dryden rewrote Shakespearean plays into heroic couplets, taming both their perceived excess and the literary exuberance that made Shakespeare Shakespeare. Linnaeus wrestled the growing swarm of plants and creatures into a *Systema Naturae*—relying on artificial and arbitrary traits, but successfully wrangling the additional data into a great chain of being. Samuel Johnson codified English into a dictionary and wrote a critique of far-flung travel with *Rasselas*, about an Ethiopian prince afflicted with a desire to travel who learns in the end that what he needed was already at home.

If not exhausted, Europe's exploring nations were fatigued. Exploration had launched to find routes to the Indies; it had succeeded, and with that success came a questioning of the need for more and a critique of what the ransacking had wrought both for the discovered peoples and those who did the discovering. In 1719 Daniel Defoe published *Robinson Crusoe*, a tale of wanderlust gone bad. After several misadventures, his father warns him what the future might bring. "My father, a wise and grave man, gave me serious and excellent counsel against what he foresaw was my design. . . . He asked me what reasons more than a mere wandering inclination" I had to set out again, and "told me it was for men of desperate fortunes on one hand, or of aspiring, superior fortunes on the other, who went abroad upon adventures, to rise by enterprise, and make themselves famous in undertakings of a nature out of the common road." The middle path was best for one such as his son. Only later generations, after enthusiasm for exploration had revived, would read the text—selectively—as a call to adventure and a triumph over adversity. More directly, and savagely, in 1726 Jonathan Swift published *Gulliver's Travels*. In 1752 Voltaire contributed *Micromegas*, a brutal take-down on the discovery of new worlds.

Gulliver is a biting satire on exploration that echoes the Old Man of Belém and Las Casas, though refracted through fantasy. Lemuel Gulliver is Crusoe as a surgeon "condemned by Nature and Fortune to an active and restless life," and so turns to the sea where, like Columbus, he undergoes four voyages of discovery that mock all that such voyages are purported to do. His first voyage leaves him shipwrecked on Lilliput, a land of tiny people northwest of Van Diemen's Land, where he becomes embroiled in civil war. The second has him marooned in Brobdingnag, a land of giants east of the Moluccas (actually the Pacific Coast of North America), where like natives brought to European capitals he is exhibited as a curiosity. The third voyage has his ship assaulted by pirates and him abandoned, like Alexander Selkirk, on an island in the Indian Ocean. He is rescued by a Floating Island, Laputa, populated by "projectors" and other visionaries, which escorts him to assorted other islands in the east, ending in Japan. He resolves not to travel again, but becomes bored and takes to sea as the captain of a merchantman. The crew mutinies, and leaves him on the first island they find while they become pirates. The island is inhabited by intelligent horses called Houyhnhnms and cretinous humans known as Yahoos. Gulliver comes to admire the former and loath the latter, but is banished back to England—a kindly Portuguese captain takes him—where he prefers the company of his horses to that of his neighbors.

Virtually every trope of the traveler's tale, the personal narrative of a voyage of discovery, and the chronicle of European exploration is present, and each is inverted. The immersion into local wars. The splendor and confusion of first contacts. The relentless misadventures of wrecks, maroonings, storms, and pirates. The use and misuse of natives. The impractical, if not scandalous, schemes for colonies, South Sea Companies, and utopian settlements. The travelogue is turned back on itself, for what Gulliver discovers are commentaries upon his own society, and a critique of geographic discovery as it was actually conducted. He is particularly disgusted with the prevailing logic by which exploration must lead to claims, and claims to conquest, and since he did not believe the lands he had visited had "a desire of being conquered and enslaved, murdered or driven out by colonies, nor abound either in gold, silver, sugar, or tobacco," he did "humbly conceive they were by no means proper objects of our zeal, our valour, or our interest." He scorned travel writing as "fables" and thought it "better, perhaps, that the adventure not have taken place, or if occurring, that its discoveries remain unknown."[1]

Later generations have tended to simplify *Gulliver* into the first voyage alone and sanitize it into a children's story or a cartoon, keeping its sense of adventures in exotic lands while stripping it of its scorn, wit, and indignation. The same has held for *Robinson Crusoe*, which in its original form is a paean to despair and a lament over the foolhardiness of wanderlust. The Great Voyages had not launched under the direction of the intelligentsia, but the era would end with their general disenchantment.

If the circumnavigations of Ferdinand Magellan and Juan Sebastián Elcano epitomized the most extraordinary and memorable of the Great Voyages, then Francis Drake and Thomas Cavendish epitomized how later circumnavigations could announce the entry of another contestant, in this case Elizabethan England. Decades later, circumnavigation found its dark side as William Dampier completed the cycle of descent into wanton wanderings as discovery devolved into privateering. His life of travel around the world looks less like voyages of exploration than the trajectories inscribed by a global pinball machine.[2]

Like the protagonists of *Robinson Crusoe* and *Gulliver's Travels*, Dampier drifted to the sea and, despite repeated failures and sufferings, kept returning. He learned the trade by sailing to Newfoundland, followed by a bout with the Royal Navy, then by voyaging to Jamaica, the West Indies, and Yucatán, where the latter crew blundered around the Caribbean. In what might stand as a metaphor on his life, Dampier wrote that "in these rambles we got as much experience as if we had been sent out on design." He bounced from one voyage to another, one job to another, but mostly beat around Spanish America, along with trips to Sierra Leone and Virginia, and then followed a succession of sails to the Pacific and Indian Oceans. Mostly the voyages were about piracy, which Dampier elevated by calling it privateering. Eventually, after 12 years, he made his way back to England. In 1697 he wrote up his experiences into *A New Voyage Round the World*, a rousing literary success. He followed with other volumes, picking through the gleanings of his memory and diary. Throughout, he recast opportunistic voyages of buccaneering into voyages of discovery.

The book was enough to have him appointed in 1698 as captain of a genuine voyage of exploration to New Holland and New Guinea. Dampier proved a successful writer of travelogues but a wretched captain, and after more misadventures he returned to face court-martial. Still, he continued to turn seaward, and in 1703 commanded a privateer that returned to England in 1707 "no richer in material wealth, and considerably poorer in reputation." The next year he repeated the exploit, this time capturing a Manila galleon of dazzling wealth, though he never realized the full bounty. He reached England in 1708 and died in 1715.

Three times he had circumnavigated the world, and another voyage took him to Australia and back—perhaps a record for the era. A good observer, a serviceable writer, he made his reputation with his *New Voyage Round the World* and some notes on hydrography and winds. He called himself an explorer rather than a pirate; his wanderings, voyages of discovery rather than the flotsam of wanderlust; his standing, that of a bold leader rather than of a dismal commander and frequent dissolute. With a few exceptions, and those the outcome of storms and disasters, he went where Europeans had already gone—that, after all, was where a privateer must go to find plunder and where, after shipwrecks or maroonings, a survivor might expect to find rescue ships in transit.

In many respects Dampier was an English Mendes Pinto, on picaresque wanderings throughout the imperium of the First Age. Others would pluck from his accounts the nucleus for travel literature (Defoe got the inspiration for *Crusoe* from the story of Alexander Selkirk, left on Más a Tierra during

Dampier's third voyage and picked up on his fourth). His four circumnavigations mattered less than the Armada de Molucca's one, and more resemble the four voyages of Lemuel Gulliver than those of Amerigo Vespucci. William Dampier was the First Age drifting out to sea under a fast ebb tide.

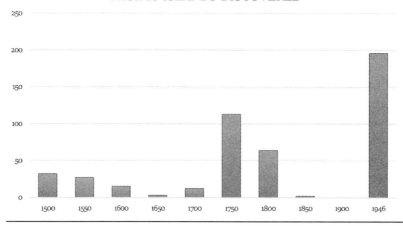

PACIFIC ISLANDS DISCOVERED

The European discovery of Pacific islands by 50-year increments. The graph illustrates nicely the three great ages. The first begins large, and then falls off as commercial trading replaces exploring. The second age rises and falls as exploration searches out new islands, which become harder to find over time. The third age marks the discovery of sea mounts—what had once been islands that subsequently eroded and sank beneath sea level. Data from Henry Menard, *Islands*.

Caravela redonda from the armada de João Serrão, an example of the hybrid rigging and hulls that characterized Iberian ships of discovery. Detail from *Livro das Armadas*, ca. 1502.

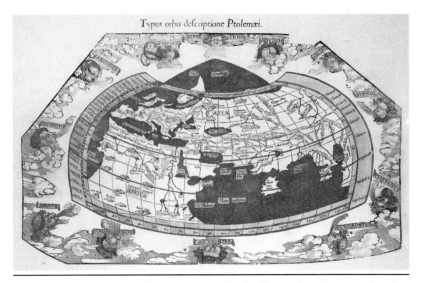

Typus orbis, descriptione Ptolemaei, written for Claudius Ptolemy's *Geographia*, first published in 150 CE; rediscovered ca. 1417; here as published in the 1541 edition. Note the curved Earth, and the use of latitude and longitude. Courtesy Library of Congress, Geography and Map Division.

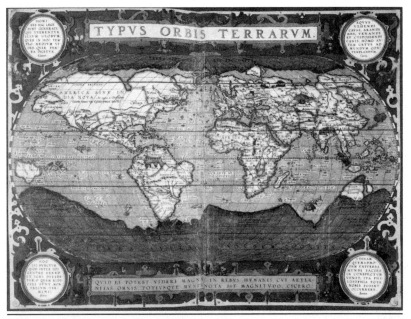

Abraham Ortelius world map, published in *Theatrum Orbis Terrarum*, regarded as the first modern atlas (1570). Published in the Netherlands, it incorporated most of the major discoveries of the First Age. Courtesy Library of Congress, Geography and Map Division.

FERDINANDES MAGALANES LVSITANVS *anfractuoso curfyo fuperato, Cos tellurì ad Auftrum nomen dedit, ciusque navis omnium prima atque noviffima Solis curfum in terris emulata, terror totius globum circumijt. An. Sal. ∞·D·XXII.*

Ferdinand Magellan passing through his eponymous strait. From Jan van der Straet, in *Americae Retectio*, ca. 1580s.

Alexander von Humboldt (left) amid the intellectual splendor of the South American rainforest, the epitome of the explorer as Romantic hero. Eduard Ender, *Alexander von Humboldt und Aimé Bonpland im Urwald* (1856).

BOOK II

Corps of Discovery

The Second Great Age of Discovery

Come, my friends,
'Tis not too late to seek a newer world. . . .
For my purpose holds
To sail beyond the sunset, and the baths
Of all the western stars, until I die.

—ALFRED TENNYSON, "ULYSSES" (1842)

So far we have acted like a couple of mad things. We pick up one thing only to drop it for another. Bonpland keeps telling me that he'll go out of his mind if the wonders don't cease soon.

—ALEXANDER VON HUMBOLDT,

ON LANDING AT CUMANÁ, VENEZUELA (1799)[1]

The Enlightenment Explores

The object of your mission is to explore the Missouri river, & such principal streams of it, as, by its course and communication with the waters of the Pacific ocean, whether the Columbia, Oregan, Colorado or any other river may offer the most direct & practicable water communication across this continent for the purposes of commerce.[1]

—PRESIDENT THOMAS JEFFERSON'S INSTRUCTIONS TO
MERIWETHER LEWIS, JUNE 20, 1803

Between 1798 and 1806 three extended expeditions announced a new era of exploration: they were for the coming century what the conquest of Ceuta, the discovery of Madeira, and the occupation of the Canaries were for the 15th century. They displayed, not unexpectedly, many characteristics with preceding voyages of discovery with which they clearly shared a common heritage and genealogy. Yet they were not the prototypes, or the heralds and annunciators, for the era; that honor belonged with the global array of astronomers attempting to track the transits of Venus, the natural philosophers who traveled to the arctic and equator to measure the Earth, and even a handful of traveling m'lords on a Grand Tour that forced them to contemplate volcanoes and natural history as well as the relics of antiquity and Old Masters.

Rather, the Institute of Egypt, the travels of Alexander von Humboldt, and the Corps of Discovery under Lewis and Clark demonstrated publicly how all the pieces—some left from the First Age, some adapted from other purposes, a few invented—could combine to sculpt a new kind of exploration.

They reflected a rekindled enthusiasm for discovery and they so rebuilt the previous models that the expeditions could seem a novel, separate creation. They were demonstrably new—in purpose, in energizing rivalries, in styles, in arenas for discovery, in geopolitical and cultural contexts, and in their sense of how the world worked and how Europe fit into it. They became the models for what would follow.

Collectively they proclaim a Second Great Age of Discovery.

In 1798 Napoleon Bonaparte invaded Egypt and established as part of the occupation the Institute of Egypt, an academy of scientists and artists who would rediscover that most enigmatic land of antiquity and describe it in more contemporary terms. It was, in one sense, a forced conversion to modernity on a people who didn't want it, and as with religious missionizing for Spain, a way of sanctioning conquest. It was a *mission civilisatrice* dressed for the Enlightenment.

The institute (or "Commission on Arts and Sciences Attached to the Army of the East") included 151 savants, many such as Gaspard Monge, Joseph Fourier, and Geoffroy Saint-Hilaire drawn from the creme of French intelligentsia. For three years they labored to inventory Egypt along the parameters of modern science, eventually consolidated into the multivolume *The Description of Egypt*, authorized in 1802 and completed in 1828 with 26 volumes of text and figures. It was the French *Encyclopédie* transported to Egypt. At the Battle of the Pyramids Napoleon's legions, with modern weapons and infantry squares, had destroyed the Mamelukes and their antiquated cavalry. The Institute of Egypt was the intellectual complement to the invasion that would, in theory, break the ancien régime culturally and wipe away millennia of obscurantism and superstition. The savants would also support the administration of the French conquest, modernizing governance along more "rational" lines.

Both projects ended badly. The French navy was obliterated in the Battle of the Nile, and its army shriveled away under a relentless insurgency until finally surrendering to Britain. Meanwhile, Napoleon had silently abandoned the campaign in order to return to France and stage a coup d'état.

The institute suffered as well. Twelve members left, 26 died in Egypt, and five died soon after returning to France—an attrition of 28 percent. Many of the survivors never fully recovered. Its best-known relic may also be its most apt symbol: the Rosetta stone was found by a French soldier before being handed over to the savants, and was then given to the British upon surrender, who deposited it in the British Museum.

Earlier in this Enlightenment century academies of science had sponsored expeditions. Now a military expedition sponsored an academy of science. The *Description of Egypt* set a new standard for the literature of discovery.

Indirectly, it inspired Alexander von Humboldt, then in Paris, aroused by revolutionary feelings and eager to explore. Forestalled from Egypt and the round-the-world voyage of Nicolas Baudin, he turned to Spanish America, then a land with its own ancient monuments in some ways as closed as Mameluke Egypt. Like Columbus and Magellan he appealed to the Spanish monarchy, who granted him a passport. Unlike Napoleon, he traveled with sextant and barometer rather than sword and musket. For five years (1799–1804) he trekked around Cuba, Venezuela, Columbia, Ecuador, and Mexico, collecting, measuring, mapping, documenting. Few areas were truly unknown: they had been recorded by conquistadores, missionaries, and in Ecuador, an earlier expedition under the Paris Academy of Science. But they were new to the eyes and instruments of modern science. Humboldt's was a vast project of rediscovery and, as so often in exploration, of translation, this time into the emerging sciences of geology, biology, and geography, of which he was a founder. For his labors he was honored as a "second Columbus," the scientific discoverer of America.

Yet he was far from the swashbuckling conquistador, ardent missionary, or gold-addled mariner that personified the Renaissance voyages. He was an explorer for the Enlightenment, one stripped of polluting bonds with politics, commerce, and religion. His ruling ambition was to create a *physique générale*, a universal synthesis, "by bringing together of all the phenomena and creations which the earth has to offer" and "of all the forms of knowledge which deal with the modifications of matter." On his departure to the New

World, he wrote of his vision, "I must find out about the unity of nature." On his return to Europe, he stopped by the young United States and dined with President Thomas Jefferson the same month Meriwether Lewis and William Clark left St. Louis with their Corps of Discovery to cross North America.[2]

That epic undertaking, the third expedition, could not stand aside from the others. It was part Enlightenment, part empire building, part scientific adventure, and for young America, a founding epic, the equivalent to the *Lusíads*.

Humboldt's long expedition provided a template for what Jefferson wished Lewis and Clark might achieve; and the uncertain borders of Louisiana—purchased from Napoleon—allowed plausible presentation of the survey as a scientific party. It recorded plenty of natural history—its journals have become iconic texts—but it was a military party, captained by officers, operating under military discipline, intending to advance a national objective, which was part of an expansionist ambition. The Louisiana Purchase doubled the size of the United States at a time when long-distance communication was by horse or sail, before telegraphs and railways. The expedition was the exploratory medium for understanding just what the United States had purchased. If the Louisiana Purchase set the conditions for further expansion, the Lewis and Clark expedition established the primary means by which it would become known.

Meriwether Lewis had served as Jefferson's personal secretary, and had received special training in natural history in Philadelphia, where America's own savants had gathered. Jefferson had made a name for himself as an early natural historian twenty years earlier with his *Notes on the State of Virginia*. Like Humboldt, he had experienced revolutionary France, understood the tremors and shakings that signified the advent of modern science as a cultural force and of democracy as a political one, and he intended that his expedition should inventory America's freshly acquired lands with the new methods suitable for a new nation, an omnium-gatherum of "useful knowledge" quite apart from determining a route across the continent and fixing the latitude and longitude of rivers, rapids, islands, and portages. In his instructions Jefferson listed examples:

The commerce which may be carried on with the people inhabiting the line you will pursue, renders a knowledge of these people important. . . . Other objects worthy of notice will be the soil & face of the country, its growth & vegetable productions, especially those not of the U.S. the animals of the country generally, & especially those not known in the U.S. the remains & accounts of any which may be deemed rare or extinct; the mineral productions of every kind; but more particularly metals, limestone, pit coal & saltpetre; salines & mineral waters, noting the temperature of the last & such circumstances as may indicate their character; volcanic appearances; climate as characterized by the thermometer, by the proportion of rainy, cloudy & clear days, by lightning, hail, snow, ice, by the access & recess of frost, by the winds, prevailing at different seasons, the dates at which particular plants put forth or lose their flowers, or leaf, times of appearance of particular birds, reptiles or insects.[3]

This was not the first foray into the plains—French and Canadian fur traders had probed as far back as the 17th century—nor the first crossing of North America—Alexander Mackenzie had accomplished that from Canada. But it was the most celebrated and significant, fundamental to America's national myth, and it can stand for all those surveys to come in the service of European settler societies. They were each, in some way, origin stories, and in America's case, less tainted by the Spanish legacy left by Columbus.

The expedition crossed the country. In the popular imagination it began in St. Louis, where it could boat up the Missouri River. But it also began in Philadelphia where Lewis studied, and in Washington, D.C., in 1803 where he and Jefferson devised his orders, and in Pittsburgh where he rafted with supplies down the Ohio River to Camp Wood, where he wintered over prior to his May 1804 departure up the Missouri. The Corps of Discovery followed the Missouri to the Rocky Mountains in western Montana, crossed at Lolo Pass, found their way to the Columbia River, and then to the Pacific Ocean. They returned in September 1806. Altogether they had traversed the continent. And true to creation epics, the two captains epitomized the genres' alloy of triumph and tragedy. William Clark settled into a career as governor of Missouri Territory, superintendent of Indian affairs, and folk hero for the republic; Meriwether Lewis, unable to cope with fame and finish his journal, ended with suicide.

The Lewis and Clark expedition became the inspiration for the Rocky Mountain fur trade and the template for dozens of expeditions that followed—the Pike expedition to the Southwest, the Long Expedition to the Great Plains, the swarm of surveys by the Army Corps of Topographical Engineers that traced America's new borders after the Mexican War, that platted wagon routes and mapped tribes, rivers, and mountains, that oversaw four grand traverses in the search for a route for a transcontinental railroad.

All looked to Humboldt for a general model and to Lewis and Clark as his American avatar. If Columbus marked the discovery of the New World, and Humboldt its scientific rediscovery, Lewis and Clark tracked the scientific discovery of its dominant democracy.

The First Age had run in loose parallel with the Renaissance, two tiles of a complex mosaic undergoing metamorphosis. Discovery had bolted away on its own, and intellectual culture trailed, trying to organize the clutter and debris and novelty of discovery into poorly suited medieval institutions and to reconcile its startling revelations with Renaissance-recovered texts. In the Second Age, intellectual culture, in the form of an Enlightenment inspired by the example of modern science, and those other disciplines and arts influenced by the power of Reason, often led.

Only occasionally was the scientific impulse sufficient by itself, but it fraternized easily with other goals. It was part of an extensive secularization that replaced missionaries with naturalists and converted religious goals to more naturalized morals and to more secular arguments for settlement. Once again, exploration became an ally, willing or not, to empire, this time with special force to the societies swarming with colonists that replaced the indigenous peoples. The role Faith had assumed in the First Age, Science assumed in the second—or rather, that transfiguration of elite culture that science had wrought into an Enlightenment. It helped redefine why exploration occurred, helped shape where and how explorers went, and helped justify its often ambivalent consequences. As Newtonian laws of nature replaced Aristotelian natural law, so the linkage of exploration with natural philosophy went from a weak van der Waals bond to a strong nuclear one.

The Second Age marked the advent of the Enlightenment as explorer.

2

Grand Tours and Great Excursions

Every blockhead does that; my Grand Tour shall be one round the whole globe.

—JOSEPH BANKS, CHIEF SCIENTIST ON CAPTAIN JAMES COOK'S FIRST CIRCUMNAVIGATION[1]

The Second Age began by completing the tasks left unfinished by the First Age. The Great Voyages had launched to find routes to known lands, and then stumbled upon new lands, and ultimately discovered that the world's seas were one; it had not set out to map the world ocean and its shore. The Second Age did. It sought out the last straits, sketched the last unchartered littorals, plotted the immense scatter of Pacific islands, and discovered the final continents.

The new circumnavigators did not simply go to new places, they saw with new eyes. They carried members of scientific academies, and often boasted artists and skilled writers. If the First Age was prepared to quest after mirages and geographic chimeras like the Fountain of Youth or the Seven Cities of Cibola, the Second was inclined to search for the fonts of the Nile or abstractions like the magnetic pole and a continent plated in ice. They wanted to advance their nations' wealth and power, but they also valued specimens of birds and beetles, the artifacts and ceremonies of tribesmen remote from civilization, data that could support or subvert scientific theories like gravity and evolution. As well as spices, the Moluccas held exotic butterflies. As well as commerce, the Atlantic isles were caravanserai for science. As well as bullion, South America held caches of bones belonging to extinct creatures.

This transition did not appear instantly. It began in pieces, scattered across many nations, created to various purposes, beholden to particular personalities. Most were one-off events; few talked one to another. They were like tiny mammals scurrying beneath the bulk of aging dinosaurs, yet when conditions changed, they would evolve rapidly into forms that would take over much of the Earth.

Whether as the last of the Great Voyages, or the first of the Second Age's, the two expeditions of Vitus Bering mark a transition. They were anomalous because they were the creation of Tsar Peter the Great, always fascinated by ships and Western enthusiasms. Cossacks and promyshlenniki had surged across Eurasia but had halted at the northern Pacific. Boats had been critical to that rapid expansion, but pirogues on rivers, not blue-water ships. Peter wanted to know the perimeter of Greater Siberia, and he chose to pursue that ambition through the kind of voyages Europeans had begun making. In 1724 he established, again on a European model, an academy of sciences; representatives from it would accompany any exploration, which would grant those surveys an intellectual patina beyond piles of pelts.

Peter died the next year, but his intentions had been set into motion, and they would continue after his passing. The party consisted of Germans, Danes, and Russians; a Dane, Vitus Bering, would command; a German member of the academy, Georg Steller, would serve as naturalist; and another, Gerhard Müller, as historian and geographer. It took the expedition two years to cross from St. Petersburg to Okhotsk. There they built a ship and acquired another, and in the summer of 1728 explored the Sea of Okhotsk, Kamchatka, and what they named the Bering Strait—the last of the great straits to be discovered, this one less valuable as the corridor to the Arctic Ocean than as evidence that Asia and America were separate. Bering returned to St. Petersburg in 1730.

Their geographic discoveries, however, left unresolved the actual dimensions of the strait. Particularly uncertain was whether its eastern shore ended with the American mainland or amid Alaskan archipelagos. The Russian state decided to complete the task. Bering was told to return to Kamchatka and confirm, while an elaborate surveying party would remap the northern shore of the Arctic. These were enormous undertakings, eventually known as the Second Kamchatka Expedition and the Great Northern Expedition, lasting ten years, involving nearly 3,000 people, and costing an estimated one-sixth of the Russian state budget. This time the expeditions included academicians like G. F. Müller, J. G. Gmelin, Louis Delisle de la Croyère, and Georg Steller; botanists, astronomers, naturalists, early ethnographers. They translated what indigenes and promyshlenniki knew into the formal algorithms of cartography and codified local knowledge and their own collections according to the science of the day.

Bering died on his return after a miserable shipwreck off the coast of Kamchatka, and his expedition ended in acrimony between its Russian and foreign members; but his two ships did sight mainland Alaska, and one, with Steller, stopped for a day at Kodiak Island. It had taken Georg Steller ten years to have ten hours in the New World. The upshot catapulted Russia to the forefront of advanced exploration, but so dependent were elaborate projects like these on the tsar that they faded as future rulers turned from them to other interests. The two Kamchatka expeditions were, like their originating source, an anomaly: they were Peter's projects, preserved in his memory, like statues in a St. Petersburg square. The experience was not institutionalized into a program. Not until the Second Age matured would Russia return to the forefront.

While the Great Northern and Kamchatka expeditions were completing their tasks, Europe's natural philosophers were embroiled in a controversy over competing theories of gravity, one by Isaac Newton and the other by Giovanni Cassini. This had more than arcane interest because modern science had replaced archaic cosmologies with contemporary ones derived from natural philosophy, for which gravity was fundamental.

The two theories predicted different shapes for the Earth. Newton argued that the planet would be flattened at the poles and widened at the equator, while Cassini predicted that the poles would elongate and the equator contract. In earlier ages scholars might have debated from books; in an age agog with the promise of science, they looked to the book of nature. The debate could be resolved by tests, not texts. Measure a length of an arc of meridian at the poles and the equator and compare the results. This required travel, and the voyage of discovery was retrofitted from exploration to experimentation.

The Paris Academy of Sciences oversaw the project, dispatching two expeditions in 1735, one north and one to the equator. The Lapland expedition was headed by Pierre de Maupertuis, a prodigy elected to the academy at 25, a proponent of Newton, and a favorite of Voltaire. He traveled to Torneå at the head of the Baltic Sea, laid down his triangles, and returned in 1737. The data supported Newton, as did a recalculation along the meridian between Paris and Amiens, and data received from the Cape of Good Hope. His triumph earned him a post at the Berlin Academy of Sciences and his touchy personality the enmity of Voltaire, who satirized the expedition and its leader mercilessly in *Micromegas*.

The Ecuador expedition, or French Geodesic Mission, had a large complement of savants. While Pierre Bouguer was the official head, the man most closely associated with it was Charles Marie de La Condamine, a former soldier, a young member of the academy and another associate of Voltaire. Spain had agreed reluctantly to allow the party into its closed lands, the local population was deeply suspicious, and the scientists had to endure a long trial before the Audiencia at Quito. "Some consider us little better than lunatics. Others impute our whole proceedings to the fact that we are endeavoring to discover some rich minerals or buried treasure," noted Antonio de Ulloa, one of the Spaniards assigned to the mission. "When we inform them of the real motive of the expedition, it causes much astonishment." Here, indeed, was a new style of explorer.[2]

The work went slowly, not completed until 1739, and the trial not until 1742. Once again, the data supported Newton. But over their long tenure, the expedition had become familiar with many features of the natural scene—wonders not visible to the eyes of Pizarros and Vargas Machucas. In fact, Spanish America was known, but not fully discovered, and certainly not integrated into the intellectual imperium of Europe. Its wonders awaited a younger generation willing to see anew.

Over their prolonged stay some members had died, and some had married and wished to remain; the rest found their way home by various means.

The prototype of the new explorer, La Condamine, chose to traverse the Andes and boat down the Amazon River. He had a map Jesuits had drawn from their missions, but he would recalibrate that best-guess cartography with exact measurements of latitude and longitude, for which he carried an 18-foot telescope. He returned to France in 1745.

But it was the ordeal of Isabella Godin des Odonais that gripped the imagination of Europe. An Ecuadorian creole, she married the expedition's mapmaker, Jean Godin. When the expedition broke up, Jean went down the Amazon to make arrangements for her to join him, but ended up in French Guiana, there waiting permission from Portugal to voyage up the river. Portugal eventually dispatched a galliot to assist. Delays mounted, years passed. Finally, in 1769, her children having died, Isabella resolved to join her husband and set out on her odyssey over the mountains and down the river. Her party met with one disaster after another; one member after another died until only she remained, wandering in the rainforest. Some natives found her, then helped her reach a mission at Loreto, where the galliot eventually met her and took her to her husband in Cayenne. Their reunion came 20 years after they had separated. They reached Paris in 1773—39 years after the French Geodesic Mission had first sailed.

By then the originating purpose of the project was long forgotten; what remained were the stories it spawned. The two expeditions made for great science; the story of Isabella Godin made for great romance of a sort much different from the pseudochivalric tales that Cortés and Pizarro had carried with them. The voyage of discovery was finding new bonds with the larger culture, no longer just a means to unveil novel lands and carry conquistadores, traders, and missionaries to them, but a means to express novel understandings of the world.

Natural philosophy found still further reasons to adapt the voyages and entradas of the previous era to its purposes. Once again astronomers led. This time the question was to determine the distance between the Earth and the Sun, which was needed to map the solar system since the ratios of planetary distances were known, but not their absolute numbers. Edmund

Halley, England's astronomer royal, noted the transit of a planet across the Sun could be triangulated to settle the question. In 1677 he observed the transit of Mercury from St. Helena. In 1691 he wrote that the transit of Venus would present ideal circumstances. In 1716 he showed how it could be done, but also explained that the transits came with an odd, infrequent rhythm. The next transits would occur in 1761 and 1769. It would be best for multiple observations from around the globe, not least to avoid the prospect of a cloudy day wiping out years of preparation.[3]

It was the opportunity of a lifetime—the collusion of celestial mechanics and the needs of natural philosophy. The Paris Academy of Science and the Royal Society of London spearheaded a campaign to rally the world's scientific community behind the project. The Royal Society's memorial to the Lords Commissioners of the Treasury lays out the "Motives" to sponsor what would surely become a complex and costly undertaking. These are "the Improvement of Astronomy and the Honour of this Nation: which seems to be more particularly concerned with the exact observation of this rare phenomenon, that was never observed but by one Englishman. And it might afford too just ground to Foreigners for reproaching this Nation in general (not inferior to any other in every branch of Learning and more especially in Astronome); if while the French King . . . and the Court of Russia" and others are sending observers around the world, England does nothing. In brief, there is an investment in astronomy, which carries a hint of commercial interest since astronomy was closely valenced with navigation (notably the vexing problem of longitude), and hence with commerce. But mostly there is an amalgam of pride and shame—pride in what England had pioneered, shame in letting others walk off with the honors.[4]

At the time Britain and France were engaged in a complex Seven Years' War that ranged from India to North America. The transit of Venus was an opportunity both to contain that conflict and to wage it on cultural grounds. It speaks volumes that the power of natural philosophy was such that warring nations might allow passage for participants. Not for all—India was too great a prize. A French frigate outside Plymouth convinced Mason and Dixon to avoid Bencoolen, while Le Gentil remained stuck in Pondicherry, then under British blockade. Still, transit was widely permitted in ways that were unthinkable to the First Age, which had barely permitted foreign missionaries to enter and had claimed closed seas as well as lands. But then science had not schismatized like Christianity, and rival disciplines and academies were less territorial than competing sects and religious orders.

The 1761 transit assembled 120 observers from Europe (106), Russia (5), and posts throughout the European imperium—Cape Colony, India, Baja California, St. Helena, Kamchatka, Sumatra, the Isle of Rodrigues, and the Gulf of Guinea. The observations traced out the trade routes of the European imperium, save for the Iberians, who harbored a deep wariness of foreign savants traveling through their closed colonies. The 1769 transit orchestrated 150 observers, fewer in Europe (92) and more in North America (30). The French dominated the first transit; the British, the second. But then Britain had won the Seven Years' War.

In the end, while the data were not definitive in establishing the "frame of the world," they came within 2 percent of modern reckoning. It was a remarkable achievement given that the sites selected were those most possible, not those most desirable, and that instruments and observers were imprecise (analogue, in today's parlance). As an exercise in exploration, the transits were spectacular successes. They showed how old modes of voyaging and trekking could be refitted to serve the ambitions of science. They advertised, if indirectly, that natural history offered as much interest as natural philosophy, for with their long waits in exotic places expedition members had begun recording and mapping the wonders on Earth along with those in the heavens. They helped redefine the ambitions of exploration, turning the wheel of secularization. And they demonstrated the possibility for international collaboration in unprecedented ways. It is impossible to imagine Britain and France, much less them, along with Portugal and Spain and Holland, agreeing in the 16th century to sponsor multiyear voyages for abstruse questions about the design of the solar system. However much petitioners might appeal to nationalism, science was proving transnational, and it would eventually nudge exploration in the same roughly transcendent directions.

Meanwhile, a pivot to natural history was underway. The movement turned on Carl von Linné, better known as Linnaeus. Linnaeus made his name as a promising naturalist by being the first European to grow a banana; as a mature naturalist, by inventing modern systematics, a method to classify all of life on Earth; and as an explorer, by a 1732 trek to Lapland. Not a closet botanist, he went out into the world to see and gather. He became the dominant figure for natural history over the 18th century and because of his excursions, a model for explorers seeking a wider world than Sweden.

The range of his influence was astounding. His *Systema Naturae* went through a dozen editions over his long lifetime—a bible for naturalists, as essential to understanding landscapes as navigation charts were for mariners. From his professorship at Uppsala he established a research garden, the model for others around the world. He organized excursions into the surrounding countryside, elaborate fests in which he would lead parties to discover, name, and collect nature's bounty. These he expanded into Lapland-like transects through the major provinces of Sweden—cross-sectional inventories of the biotic landscape, six traverses in all. It was on his example that Maupertuis chose Torneå as his base camp. It was from his excursions that natural-history inventories became a staple of Second-Age expeditions.

His fame rang throughout Europe, which drew students. Twelve of them—his apostles—projected his methods around the European imperium. From 1745 to 1799 they traveled with merchantmen (often hiring as physicians) or where possible with formal expeditions to British North America, Brazil, Surinam, the Windward Islands, West Africa, around the Mediterranean and Middle East, Russia, India, Sri Lanka, South Africa, New Zealand, Australia, Malacca, China, Japan, and even skirting the Antarctic pack ice. Much to Linnaeus's grief, six died on their travels, secular martyrs to science.

There was a practical purpose behind his longer excursions (Lapland was a youthful bildungsroman). He reported on useful plants, the character of ores, the quality of soils, the prospects for mines and agriculture. But the sheer delight in nature and its prodigal abundance was always there, too. His apostles were emissaries of science, not of a state. They were not the foragers and scouts for new conquest: they went where European posts were already well established, seeking to collect plants and insects, as many as possible, not hoard spices and silver. What they gathered went into libraries and research gardens, not treasuries or houses of trade. What they learned they published as herbaria.

But if they sorted through what seemed to many the vignettes and left-overs of landscapes, if they swept up the chads and wastes that no one wanted, and could present a sometimes comic appearance, with their nets and lenses, compared with hard-bitten captains, they helped turn explora-tion toward new directions, and unlike the astronomers, theirs was not a one-time spectacle. Birds and beetles were everywhere. By the time Linnaeus passed away, no self-respecting corps of discovery took to the field without its resident naturalist.

Throughout the 18th century British aristocrats adopted a more humanistic mode of travel, one complete with an entourage that included a governor, tutors, accountant, servants, and guidebook. The reasons were several, but the greatest may have been the slough of lethargy into which the British uni-versity had sunk. Edward Gibbon declared his time at Oxford as "the most idle and unprofitable" of his life. The sentiment became almost universal that only travel outside the country could salvage the education an aristocrat and gentleman required (even Adam Smith thought so). The journey to Europe became an expected rite of passage.[5]

The anchor points were Paris and Rome, where young m'lords could view and copy the masterpieces of art and reread and contemplate the fate of ancient empires. Since travel was overland, the Netherlands and Germany were added to France and Italy. Gradually, the routes congealed, and the itin-erary became a curriculum. In 1749 the ritual had progressed far enough that Thomas Nugent, the Baedeker of his age, published an authoritative guide-book in four volumes; in 1756 he rewrote it. "The whole," he noted with false modesty, "if I mistake not, [is] what is commonly called the grand tour."[6]

Nugent begins by citing works of the ancients as justification, noting that "travelling, even in the remotest ages, was reckoned so useful a custom, as to be judged the only means of improving the understanding, and of acquir-ing a high degree of reputation." Moreover, "agreeably to this it has been observed, that those who first distinguished themselves in the republic of letters, were all travellers, who owed their learning, name and reputation to different peregrinations." Not least, the Great Voyages from Columbus to "the great Lord Anson" had revived the tradition. Here was an activity in which moderns could emulate the ancients. They could reread classic texts, experience travel to foreign lands, and visit the decayed splendor of antiquity. They could, in brief, translate an ancient practice into modern forms. The Grand Tour had the sanction of antiquity (Nugent even quoted in Greek); was universally applauded as a means to knowledge and stature; and could promote a literary reputation.[7]

This was hardly of a piece with raw exploration in new lands. But it conditioned the British nobility to the value of visiting new lands, and it bonded travel with art and literature. This was the Linnaean excursion outfitted with painters, writers, and reluctant classicists rather than botanists, and it toured museums, the ruins of antiquity, and European capitals rather than barren Atlantic isles, Mayan ruins, or Tenochtitlán, and while travel over the Alps might be by coach and sedan, it *was* travel. It forced the grand tourists to look at Etna and Vesuvius, to climb mountains and cross rivers, to poke through Pompey and sit on the steps of the capitol at Rome, to meet odd peoples and strange customs, to collect curiosities as souvenirs, to record the sights with pen and paint, and to write up the experience.

It got skilled artists to document the sights—emulating art was part of the educational goals, after all. But some grand tourists included artists as well, or hired artists after their return. Eventually expeditions adopted the practice, producing a visual record beyond what naval officers trained as draftsmen could produce. Art, too, was changing in ways that complemented exploration. The Grand Manner morphed from the scenes of antiquity to modern set-pieces, such as Benjamin West's *Death of Wolfe*, that, in turn, could inspire paintings like C. J. L. Portman's *Death of Willem Barents* that mimic the same composition. Even more, the great themes of human history could segue into those of natural history. By the time the Second Age was reaching a climax, landscape was in vogue, artists were sought out or sought out for themselves expeditions to which they could attach themselves, and some of the most formidable records of the era's exploring are those memorialized in paint.

In a similar way the Grand Tour pushed the travelogue into new territory, not simply an annotated ship's log but something with a genuine literary sensibility, a narrative that might complement the novel, a form then emerging. Such writing went far beyond guidebooks. In 1765 Tobias Smollett published *Travels Through France and Italy*. A full-bodied account of the lands he passed through, it resembled a natural history (and ethnography), full of telling details and character sketches. Unfortunately, Smollett was still mourning the loss of his only child, and his account was as splenetic as it was lively. Laurence Sterne rebutted his gloom with *A Sentimental Journey Through France and Italy*, written while he was ailing (and published three weeks before his death in 1768).

The Grand Tour had become a venue through which two of the great literary figures of the age could remake an old genre. Explorers further adapted the style (or commissioned ghost writers to do it for them) as they encountered horizons far beyond Italy's melancholy ruins and the small villages of France. The old "voyages 'round the world" had gripped the imaginations of elites. The new voyages, retrofitted to accommodate genres like the novel much as they had learned to accommodate naturalists, could fire the imaginations of a literate middle class as well. The personal narrative of an expedition became one of the most popular species of literature throughout the long tenure of the Second Age. John Black, English translator of Humboldt's *Political Essay on the Kingdom of New Spain*, agreed with Bernardin de St. Pierre that "by far the most valuable and entertaining part of modern literature is the department filled up by travellers." While the ancients had known only a "small circle" of the world, today "there is hardly a nook in the most remote corner of the world of which we do not now possess some description, and with the inhabitants of which we are not more or less acquainted." So popular was the genre that Black "regretted" that "almost nothing is so very insipid that will not be devoured in the shape of travels." Like specimens of beetles, books overflowed their former bounds.[8]

These shards of a new style of exploration demanded a powerful force to pull them together. It required a geopolitical rivalry as acute as that between Spain and Portugal that had driven the First Age. It found it in a revived competition between Britain and France that would play out in India, North America, the Caribbean, and the Pacific Ocean, and ultimately every place Europe could reach.

It began along existing lines—the colonies and trade routes picked off from the Iberians. Then it broadened to include the colonies in North America, the sugar islands of the West Indies, India (the great prize), and the Pacific, where a revival of circumnavigation led to the discovery of new shorelines, new islands, and new treasures to fight over. The first Hundred Years' War had spread across France; this second, thanks to three centuries of discovery, sprawled across the globe. The Seven Years' War (1756–63) is commonly regarded as the first world war.

The renewal of circumnavigation focused on the Pacific. The pivot may well be the famous voyage of George Anson during the War of Jenkin's Ear. Anson was a veteran officer of the Royal Navy who in 1737 was sent to the Pacific with a squadron to harass Spanish possessions and shipping. The climax came in 1743 when—like Drake and Cavendish before him—he captured the Manila galleon, which brought him wealth and fame, along with Spanish charts of the Pacific, eventually landing him in Parliament and making him Lord of the Admiralty. By then, Britain was at war with France. In 1748 Anson published the memoir of his journey, *Voyage Round the World*, and to wealth and power he added literary fame. Like Drake's before him, the book inspired several generations of Britons.

A fundamental realignment of geopolitics was underway. Between them Britain and France commenced a rediscovery of the Pacific. By now Dutch energies were receding, after Abel Tasman's voyages around New Holland and New Zealand, and Jacob Roggeveen's discoveries of Easter Island and Samoa. The new wave, for Britain, included John Byron, who had sailed under Anson, Samuel Wallis, who found Tahiti, and James Cook; and for France, Louis de Bougainville and Jean-François de La Pérouse, who vanished, and Antoine La D'Entrecasteau, who sailed to find him. They bumped into new islands, and then they searched for them. The Pacific isle came to have the standing for this new era of circumnavigation as Atlantic isles had before. Tahiti and Hawai'i had become the Canaries and Madeira of the Second Age. The intensity of the outburst even inspired Spain, the country most knowledgeable about the Pacific, to reclaim its status by sponsoring a modern-style expedition under Alessandro Malaspina. But Spain soon shed its brief fling with the Enlightenment and then lost its major American colonies amid the shocks of the Napoleonic wars.

The rivalry between France and Britain—commercial, military, cultural—went global. The Seven Years' War segued into the American Revolution, which France joined. Then Britain and France squared off for the Napoleonic Wars. After France receded, other European nations stepped in, as they had when the Iberian rivalry had cooled centuries before. But that mid-18th-century face-off was the spark that started an era, and it found tinder in the kindling from other aspects of the culture that took a renewed interest in travel and scientific discovery. George Anson had sailed to disrupt Spain's Pacific imperium, and when he discovered in the Acapulco galleon, plying between Mexico and the Philippines, a cache of Spanish charts for Pacific islands, some accurate, some fantasy, they were incidental to the ship's hold of bullion. James Cook sailed to measure the transit of Venus, and finding new islands was core to his mission.

The rivalry set off a conflict, hot wars and cold ones, that moved from the world's coastlines to its continental interiors. With few exceptions, notably northeastern North America and Mexico, the First Age had colonized with trading posts and military forts, needing to work with local communities, seeking to take over regional trade (and redirect a fraction to Europe), and ruling indirectly, claiming only the top caste. There had been some emigration, enough to worry the state, but empire meant rule, not settlement. The major exceptions were the British and French American colonies, and Caribbean islands repopulated by a few European migrants and many African slaves.

In the late 18th century that pattern changed. Northern European nations became the primary imperialists, as they were the dominant explorers, and settlement by immigration and local population growth pushed inland. Exploration acquired secondary centers in these Neo-Europes. The explorer assumed an added gloss and stature as a Moses figure, leading to promised lands. Behind the movement lay an altered dynamics and definitions of empire.

The imperial contest gave momentum to exploration as a vehicle for continuing politics, trade, and war by other means. From the late 18th century to the early 20th most exploration meant either Britain, its imperial surrogates, or their competitors. The scale of the contest not only involved most of the world, extending even to Antarctica, but engaged much of the culture as well.

Science, art, literature, diplomacy—the new corps of discovery touched them all. More and more, exploration looked like its sustaining society. The Portuguese paradigm reincarnated in the guise of Victorian Britain and its rivals.

All the pieces, like molecules in a supersaturated solution, crystallized out with the three voyages of Captain James Cook. It was as though geographic discovery underwent a phase change; what had been vapor became water, with the same elements but in new forms.

Cook's first voyage illustrates nearly all the age's defining features. It was designed as a circumnavigation, not accidentally but intentionally. Its primary goal was to measure the transit of Venus on newly discovered Tahiti, then to look for unknown islands, survey the eastern coast of New Holland (Australia), and investigate New Zealand, at that point known only from one abortive landing. The mission was outfitted for science, not only the astronomical purposes needed for measuring the transit, but an all-purpose survey of natural history.

A key figure was Joseph Banks, an aristocrat keenly interested in natural history who boldly declared that "any blockhead" could do a Grand Tour; his would be a voyage 'round the world. Cook had retrofitted the HMS *Endeavour*, a Newcastle coaler, to the needs of a voyage of discovery. For the expedition's primary mission the ship's company included Charles Green, assistant to the astronomer royal. At his own expense Banks further modified the ship to accommodate his entourage, with four servants and a support staff of two naturalists, two artists, Sydney Parkinson and Alexander Buchan, a scientific secretary, and a workspace for collecting and analyzing; the naturalists were Daniel Solander and Herman Spöring, both Linnaean students. At Tahiti the expedition acquired a native man, Tupaia, who served as guide and interpreter and identified some 74 other islands as they sailed throughout Polynesia.

Upon his triumphal return Banks ascended the heights of British science, heading the Royal Society and in other ways supporting exploration (he was a founding member of the African Society, dedicated to promoting exploration in that continent). It was on his recommendation, moreover, that Britain began shipping convicts to found the colonization of Australia (since the American Revolution they could no longer be sent to North America; and besides, a settlement, even a penal colony, could help fend off French ambitions in the region). The consequences from that inaugural voyage cascaded through the chronicle of British exploration for decades.

The second voyage had as its intentions to decide whether or not a polar *Terra Australis* existed. Cook didn't find mainland Antarctica, but he circumnavigated the Southern Sea and determined, on the basis of immense icebergs ("ice mountains") that there must be a major landmass around the southern pole. The complement of savants included William Hodges, an artist; Reinhold and Georg Forster, father and son, both naturalists; along with another of Linnaeus's apostles, Anders Sparrman. Georg's personal narrative, *A Voyage Around the World*, became an international seller and helped confirm the new style of travelogue. Later, he became a friend and inspiration for the young Alexander von Humboldt.

The third voyage turned to the Pacific of North America, tracing coastlines not yet mapped. It ended in Cook's tragic death in Hawai'i. From the onset Cook had earned a reputation as someone who could meet natives fairly and peacefully. It helped that so many of the islands visited were Polynesian, so that one language could be understood throughout (much as Columbus could sail the Caribbean). Tupaia served that role, and Banks eventually picked up enough to interpret as well. The usual away party consisted of Cook, William Monkhouse, Banks, and an armed escort; captain, surgeon, chief scientist—the same composition as favored two centuries later by *Star Trek* (Captain James Cook is reportedly the model for Captain James Kirk). But something went amiss when a storm drove him back to Hawai'i after he had apparently left for good. Tensions rose, and when Cook led an armed party to shore, he was overwhelmed and killed. He was immediately honored as a martyr to science. There is even an engraving, *The Apotheosis of Captain Cook*, by P. J. de Loutherbourg from a drawing by John Webber, the ship's artist, that has him ascending heavenward, a secular assumption, with a trumpeting cherubim on one side and Athena on the other.

The contrast with Magellan, who fell in the surf at Cebu, is striking. Magellan had meddled in local politics, proselytized and converted, and died in an unnecessary battle against native enemies of his host people. Cook had left local social structures intact, engaged in no military alliances, and made no attempt at conversion. His death seemed mysterious: irritants that in early voyages had been smoothed over got out of hand and it is likely that his death was the outcome of internal quarrels among his hosts. In contrast to Dampier's three voyages, Cook's were explicitly about discovering, mapping, collecting, and establishing a formal imperial presence, however light; there was no plundering of settlements, no prowling for a Manila galleon, no mutinies or marauding. In the end his voyages completed the mapping of the littoral of the world sea, save for the Arctic shores of North America and Antarctica, the bulk of which were inaccessible by wooden sailing ships.

Cook's voyages galvanized public enthusiasms, seeping even into the consciousness of such Augustans as Samuel Johnson, who had critiqued travel in a novel of Abyssinia (*Rasselas*) as a distraction from the genuine needs of heart and mind and for whom a trip to the Isle of Skye was the journey of a lifetime and might as well have been a voyage to the Moon. When his companion and memoirist, James Boswell, informed him that he (Boswell) had found a copy of Johnson's translation of Father Jerome Lobo's travels to Ethiopia, which he had published and which became the basis for *Rasselas*, Johnson dismissed the discovery and was content to have it forgotten. Yet two years later, April 2, 1776, Boswell recorded a meeting he had with Captain James Cook, recently returned from his second voyage. Johnson was "much pleased with the conscientious accuracy of that celebrated circumnavigator." Boswell then burst out that "while I was with the Captain, I catched the enthusiasm of curiosity and adventure, and felt a strong inclination to go with him on his next voyage." It was easy, he continued, to be "carried away with the general grand and indistinct notion of a Voyage Round the World." The august doctor agreed, then dismissed the sentiment when he considered "how very little" one "can learn from such voyages."[9]

The larger culture of Enlightenment Europe, elites and folk alike, agreed with Boswell. Decades later John Stuart Mill, who could dominate intellectual discourse in his day as Samuel Johnson had in his, confessed that the "two books which I never wearied of reading were Anson's *Voyage*, so delightful to most young persons, and a collection (Hawkesworth's I believe) of voyages round the world in four volumes, beginning with Drake and ending with Cook and Bougainville." He was not alone.[10]

3

Motives and Motivators

To what purpose could a portion of our naval force be, at any time, but more especially in time of profound peace, more honourably or more usefully employed than in completing those details of geographical and hydrographical science of which the grand outlines have been boldly and broadly sketched by Cook, Vancouver and Flinders, and others of our countrymen?[1]

—JOHN BARROW, SECOND SECRETARY TO THE BRITISH ADMIRALTY

Exploration in the First Age was hardly a planned project, with timetables and expectations, like a NASA program in recent times. But neither was it an anarchic sprawl. Shipborne exploration was not cheap, and although some commercial consortia (like Bristol merchants) might sponsor a foray or two, most expeditions came with the funding and sanction of the state, or as proxies of the state.

The options were many. Royal chartered companies were one venue— Dutch, British, and French East India companies; Hudson's Bay Company; the Muscovy Company; and the like. Approved religious orders such as the Dominicans and Jesuits were another. There were plenty of freelance would-be conquistadores in Spanish America, prepared to merge plunder, war, and exploration, but the state had a vested interest in reigning them in. Their British and Dutch rivals might turn to privateering, but these, too, came with at least nominal state sponsorship and the expectation that they would pay for themselves, and contribute to the treasury as well. Moreover, states took an interest in exploration generally, beyond what they might sponsor, not only as a source of revenue and for geopolitical positioning but because free-booting nationals might provoke conflict.

The Second Age was even more diverse: exploration was more a swarm than a column. Land-based exploration, particularly from settler societies, could be done much more economically than sending ships around the world; militaries staffed expeditions, or assisted with logistical support; and a thriving civil society of wealthy patrons like Joseph Banks, industrialists like William Beardmore, scientific groups like the Royal Geographic Society, and even newspapers like the *New York Herald* encouraged states or even sponsored expeditions on their own.

As European imperialism reached its apogee, many of the instruments of empire could equally double as instruments of exploration—and did. Often they were the same, with discoverers serving as scouts for imperial ambitions. It is estimated that American exploration, including the processing of discoveries, absorbed, at times, as much as a quarter to a third of the federal budget between 1840 and 1860.[2]

These were features common to all the era's participants. But by way of illustration, consider the case of Great Britain, which was not only the age's most active protagonist, but also the one that ranged the most widely with the greatest variety of agents and sponsors.

Begin with the state. The Royal Navy, through Anson, Cook, and others, helped pioneer the new era with circumnavigations, then, once the Napoleonic Wars ended, found itself with a massive fleet and a roster of officers with little to do. John Barrow, second secretary to the Admiralty and member of the Royal Society, exploited the situation to promote exploration to Africa and most spectacularly in the Arctic, and then to the Antarctic. As with Barrow, military needs, navy ships, and scientific interest converged seamlessly along with a sense of inherited tradition. Charles Darwin's voyage of discovery around the world occurred in the HMS *Beagle*; Thomas Huxley, "Darwin's bulldog," echoed his travels on the HMS *Rattlesnake*. The expense of such endeavors could be high, and few institutions outside the military could afford the cost unaided. When the Royal Society wanted an oceano-graphic survey, the Royal Navy lent the HMS *Challenger*.

Similarly, the British army had its cadre of exploring officers, often little distinguished from spies, frequently masquerading as naturalists, who added to useful geographic knowledge. Even when not on active duty, officers could be seconded for expeditions. The Great Game across central Asia is the best-known expression, and the Indian pundits like Nain Singh and Sarat Chandra Das who breached Tibet, its most exotic proponents.

The British Empire excelled in indirect rule—some 6,000 Britons governed an empire of 250 million Indians. One technique was to operate through chartered companies. The exploration of Canada was largely the work of rival fur companies, which eventually merged into a greater Hudson's Bay Company in 1821. The British East Company could act as a surrogate state and led the exploration of India; it also furnished travel for a host of exploring naturalists from Linnaean apostles to Joseph Hooker. Less commercial and overtly political, scientific societies became avid promoters of geographic discovery—the Royal Society of London and the Royal Geographic Society (founded in 1830)—or venues for announcing discoveries, such as the Linnaean Society (established in 1788), or for housing the collections amassed by expeditions, such as the British Museum, which had its building donated by George II in 1759 and which received its founding artifacts from Hans Sloane, a Royal Society member who amassed extensive natural-history collections, including a rich cache from Jamaica. An African Society for Promoting the Discovery of the Interior Parts of Africa dispatched explorers to West Africa (Mungo Park, among them). The London Missionary Society nominally oversaw David Livingstone on his eccentric treks, which yielded some breakthrough geographic insights but only a single baptism.

And there were plenty of individuals ready to sponsor or engage in travel to unknown parts. Henry Bates and Alfred Wallace paid for their Amazonian travels by collecting specimens for wealthy patrons and museums. James Bruce combined commerce and curiosity to lead him to Ethiopia and the source of the Blue Nile. Lady Jane Franklin's appeals to find her missing husband seemingly launched a thousand ships, both officially and privately (and which achieved far more in geographic discovery than the Franklin expedition ever could have.) Samuel Baker explored the upper Nile within the context of hunting, and adventure generally. Ernest Shackleton's celebrated Antarctic expeditions had wealthy patrons as funders (which is how the Beardmore Glacier got its name).

The motives to explore were abundant, and there were plenty of people primed to be motivated. Still, there were reasons to resist. The U.S. Exploring Expedition, America's bid to use circumnavigation as a means of announcing its arrival on the world stage, could not find a senior officer willing to command it, so toxic were the politics. And Alexander Vidal, the intended captain for the HMS *Rattlesnake*, destined to explore Australia and New Guinea, declined when his wife died and left him with a young family to raise—surely, the right choice. The subsequent voyage made the reputation of Thomas Huxley, its surgeon-cum-naturalist.

But most explorers were driven, obsessive, not swayed by the logic and sentiments of quotidian life or the frustrations and trammels that rule routine decisions. Consider the case of Tadeo Haënke, a 28-year-old Bohemian naturalist whom the Austrian authorities permitted to serve on the Malaspina Expedition. An adventurer and polymath, Haënke had traveled well, spoke five languages, and had a doctorate from the University of Prague. But transportation problems meant he arrived in Cádiz two hours after the *Atrevida* and *Descubierto* had sailed. Seeking to join the expedition en route, he found a ship to take him to Buenos Aires, only to have it wreck near Montevideo. "With a copy of Linnaeus in his nightcap, he swam ashore sans baggage and instruments, only to miss the expedition once again." Undeterred, he enlisted native guides and trekked over the Andes to Chile. Along the way he took extensive notes, collected 2,500 specimens, and finally rendezvoused with Malaspina at Santiago. Less ruthless than the great captains of the First Age, Tadeo Haënke was not the most extreme specimen of the age.[3]

Consider under what contexts—public, private, personal—two of the best-known British explorers of the latter 19th century went. Expelled from Trinity College, Richard Francis Burton enlisted in the army of the East India Company, became one of the outstanding linguists of his age, learned to immerse himself in other cultures, and had a restlessness almost without peer ("the Devil drives me," he once confessed). Under the auspices of the Royal Geographic Society (RGS), he disguised himself and made a pilgrimage to Mecca; then, again with RGS sponsorship, he joined John Speke and others on treks around east Africa, and in 1856 to the African Great Lakes region. He joined the Foreign Service, was posted to Fernando Pó, from where he explored the Congo River and the coast of West Africa, and later toured Brazil and Paraguay. In 1863 he co-founded the Anthropological Society of London to further the study of cultures—few men had experienced as many as he. Subsequent appointments took him to Brazil, where he ambled across the highlands and Paraguay; to Damascus; and finally to Trieste along the Adriatic coast. Institutional cover was light, but it made him more than a simple adventurer, his many books granted him standing as an author, and his contribution to learned societies made his accounts something more than Victorian travelers' tales.

Henry Morton Stanley's life is too improbable to distill neatly; for most of it he had no official standing or establishment sponsorship. He was abandoned as an infant, grew up as a "workhouse brat," made his way to the United States, where he was adopted but not officially and so was abandoned again when his surrogate father died, changed nationality from Britain to America and back again, enlisted in both sides of the American Civil War, and until late in life remained socially unmoored. He found a career as a correspondent for the *New York Herald*, reporting on the American West, the Ottoman Empire, the British expeditionary force to Ethiopia, and the Spanish civil war. In 1869 the paper allowed him to expand his remit and search for David Livingstone, whom he found in 1871 at Ujiji on the shore of Lake Tanganyika. The *Herald* and Britain's *Daily Telegraph* then sent Stanley back to Africa to complete the great task Livingstone had set for himself, to work out the hydrography of central Africa's great lakes and rivers. From 1874 to 1877 Stanley successfully traversed the continent, from Zanzibar to the mouth of the Congo River, deciphering the immense arc of the Congo and the peculiar watersheds of the lakes. Next, Leopold II of Belgium hired Stanley to help him develop the region. One of the truly vile characters of the era, Leopold ruled the claimed lands as a private fiefdom, not as an arm of the state, and he disguised the project under a bogus society, the International

African Association, using putative scientific purposes to hide a brutal commercial enterprise, which Stanley ultimately denounced as "moral miasma." Stanley's last traverse occurred when, in 1886, he was hired to lead an armed expedition to south Sudan to rescue Emin Pasha. This nominally humanitarian mission was a nightmare march, which managed to extract Pasha, only to have him fall to his death from a balcony in Zanzibar. After Stanley returned, his life suddenly flipped. He married, stood for Parliament, received honors, and enjoyed success as a lecturer and author. Such was the public power of exploration as an ideal that an abandoned child might use it to rise to the heights of society.

The Emin Pasha Relief Expedition shows the curious mingling of motives that could underwrite exploration. It was financed by Sir William Mackinnon, a wealthy Scot, friend of Stanley, and head of the Imperial British East Africa Company, which sought to expand Britain's sphere of influence. The company could function as an imperial factor much as Hudson's Bay Company or the British East India Company had in earlier times. That a British company sponsored an expedition led by a British subject to rescue a nominally British agent could be used to establish claims against French and perhaps German rivals. Eventually the company was replaced outright by a British protectorate.

Yet the enterprise underscored the curious collusion that could occur and the consequences if private practices left messes that states had to clean up (as Belgium had to do with Leopold II's machinations in the Congo). If nominally private expeditions could finesse around public confrontations, they could just as well drag states into conflicts they had no interest in. In the case of the Emin Pasha affair, the advance guard had to fight to reach Equatoria, and the rear guard disintegrated into practices out of *Heart of Darkness*.

In 1891 the *Forum*, an American periodical, raised the core questions about the potentially unholy alliance between adventuring, freebooting, exploration, and geopolitics. "The expedition for the rescue of Emin Pasha must always remain," it editorialized, "one of the greatest feats of courage and endurance in the annals of adventure." Then the punchline:

From whom did he [Stanley] get authority to begin the series of military operations that ended in depositing Emin Pasha at Zanzibar? Under whose order
did he enlist troops and exercise among Africans the power of a general in
the field? . . . Neither the British nor the Egyptian government would pay to
send Stanley to do what the British public wanted—rescue Emin Pasha. But a
committee and the loose loan of Stanley by the King of the Belgians could not
confer authority. . . . Every lawful military enterprise has a government behind
it, to which its officers are accountable, to which they are obliged to make
careful reports. . . . No judicial machinery now exists for the investigation
of the charges which Mr Stanley brings against his officers of the rear-guard.

The expedition might, in legal eyes, seem little better than piracy—
privateering under the auspices of capitalists, with a fig leaf of geographic
discovery to cover the embarrassment. But powerful states might teeter on
the edge of war because such events could cause ripples that magnified into
political tsunamis.[4]

The Second Age was so robust because it engaged so much of its sustaining
society. Anyone who wanted, it seemed, could go exploring, reap rewards,
and ignore consequences. The American John Ledyard decided to walk
around the world only to find himself arrested in Russia. Another American
John Lloyd Stephens discovered the massive Mayan ruins of Yucatan and
bought the city of Copán for $50. Amateur archaeologists streamed through
the Holy Land. Travel books and the personal narratives of explorers were
best sellers. Henry Thoreau read them by the score in his shack at Walden
Pond.[5]

Here was a full-spectrum fusion of exploration with culture. It boasted
exploring expeditions, exploring scientists, exploring artists, exploring writers, exploring politics, exploring commerce. Explorers could be honored
as heroes; exploration helped sanitize otherwise grim realities of forced
trade and imperial rule. Explorers headed learned societies and government bureaus; they entered Parliament and became colonial governors.
Exploration made the careers of such figures as Alfred Russel Wallace, the

co-discoverer of evolution by natural selection; John Forrest, later the first premier of Western Australia; and John Wesley Powell, director of the U.S. Geological Survey and Bureau of American Ethnology. In ways hard to recapture in contemporary times, explorers had a call on public attention, if not the public purse. They were the antennae and scouts of a West keen to learn about the Earth in its remotest reaches and in all its immense biotic and cultural riches.

4

Something Old, Something New

There are many reasons which send men to the Poles, and the Intellectual Force uses them all. But the desire for knowledge for its own sake is the one which really counts. . . . Exploration is the physical expression of the Intellectual Passion.[1]

—APSLEY CHERRY-GARRARD, *THE WORST JOURNEY IN THE WORLD* (1922)

What did discovery mean in the Second Age?

In the popular imagination it means the revelation of new lands and peoples, putting into mind and onto map what had not been known before. The Second Age did this: it revealed for the West hundreds of new islands, finished mapping the world ocean, discovered two new continents, and completed the unveiling of three more.

Once again, islands make a useful index. New islands in the Atlantic had been half the origin of the First Age, with continental coasting along Africa the second; those islands then became critical ports of call for further exploration. But discovery could be confusing because many of islands were being rediscovered, often more than once, and many that were sought after were mythical. It was not until contact led to occupation, which brought them into a network of continuing exploration, that their discovery might be thought final.

The Pacific islands were different. Outside the realm of the Indies isles, the archipelago of the Spice Islands specifically, they were utterly unknown to Europeans. Each discovery was novel: each was distinctive. And there were hundreds of them scattered across the largest geographic feature on the planet. The First Age found many—77 in all—as explorers sought a westerly course to the Indies, while winds, currents, and storms blew ships hither and yon. Then the trade routes stabilized, and the number of new discoveries dropped precipitously. Voyagers stopped looking for more. In the mid-18th century, as circumnavigation renewed and competition among northern European powers sharpened, new isles were found. By 1800 some 125 previously unknown islands were mapped. By 1850 there were 64 more, in another exponential decay. The last island, Midway, was found in 1859. In roughly a century Europe had discovered, for the first time, some 191 islands.[2]

A similar pattern emerged for the continents. Australia's (New Holland's) eastern coast was explored in 1769. North America did not have its northwest littoral mapped until the late 18th century. Antarctica was identified as a continent in the early 1840s, though its full perimeter would require another century before it was reduced to cartographic satisfaction. The big challenges were their continental interiors. These were truly, for Europe, new lands, and the story of their exploration fills much of the grand narrative of the Second Age.

There were other continents, however, that were broadly understood, at least in outline. The interiors of Asia and Africa were terra incognita and begged to be explored. For Asia the process was a mix of new discovery and rediscovery. For Africa the Second Age would do the hard work of penetrating beyond the coast. And there was Europe, which was surely known to the last acre by the time the Second Age geared up for exotic places. In fact, the home continent was mostly unknown to the sensibilities of the Enlightenment and had to be explored like the rest of the world. In many ways Europe served as the proving ground for the Second Age as artists and writers on the Grand Tour and scientists measuring and mapping with mathematical rigor rediscovered or reimagined Europe into a baseline for the triangulation of other continents. What the interior seas and their Atlantic extensions did for the First Age, continental Europe did for the Second.

A hundred years after James Cook's three voyages, cross-sectional traverses had been conducted over all the continents. The exception—always the exception—was Antarctica, which resisted a full crossing. Perhaps the last hurrah of the Second Age, Ernest Shackleton's *Endurance* expedition in 1914, had just that purpose.

Discovery in the Second Age was also a deeply revisionist project. Much of its energies went into resurveys of lands already known or experienced by the First Age. The conquistador Francisco de Orellana had boated down the Amazon in 1542; the Jesuit missionary Samuel Fritz had a comprehensive map of the river system and its indigenes, a copy of which he gave to La Condamine. What La Condamine did was to resurvey and recode that knowledge into the language and concepts of modern science. And that, writ large, was what the Second Age did overall. It found genuinely new places, but even more it saw familiar scenes through new eyes. It rediscovered lands it had previously explored or even lived on. The character of discovery changed because the culture of the discovers had.

What, after all, could it mean to "discover" India or China? It meant one thing in the First Age, which sought commercial routes and factories. Europe needed to know what goods would trade, what the local politics were like, what military capabilities rivals had, and what the religious inclinations of the indigenes were. It meant something else when the sensibilities of the Enlightenment were carried on East Indiamen.

Suddenly, India appeared as a civilization rich in relics, cultural significance, and geologic features unlike those in central Europe. It held ruins to rival those of Egypt and Greece. It boasted the highest mountain range on Earth. It teemed with tribes and tongues. The exploration of India resembled the (new-era) exploration of Europe, which was progressing at the same time. A place that had been known to Europe since ancient times—Alexander the Great had fought in the Punjab, Christian missionaries had appeared in the era of the Roman Empire—and that had been an object of the Great Voyages and a place of trading factories and the governing seat of the Estado da Índia was being explored as if for the first time.

Here was a place for the scholar to become an explorer. While military engineers surveyed the coast and vital roads, projecting onto maps laid out in Cartesian grid, the prototypical explorer was William Jones, a philologist and jurist with the East India Company in Bengal, who set about interpreting Indian civilization to Enlightenment Europe by founding an Asiatic Society, by translating the Vedas, and by pioneering work on how Hindi related to some European languages, and so sketching out the genealogy of the Indo-European language group. But the syncretic enigma that was India intrigued many from amateur archaeologists like James Alexander who "discovered" the sculpted caves at Ajanta, to ethnologists like Colin Mackenzie gathering royal genealogies, folk stories, and legends much as the Brothers Grimm were in Europe, and of course to waves of naturalists like Joseph Hooker leading botanical excursions across the Himalayan foothills. Laying down the triangles that would allow for the trigonometric survey of India—the Great Arc that would permit India to be mapped according to European standards, the project that would bring mathematical modernity to a somnolent society—was a very different undertaking than Alexander Mackenzie crossing western North America or Edward Eyre trekking across the Nullarbor Plain of Australia, but they all shared a common urge to find, interpret, and where possible incorporate the whole of the Earth within Europe's intellectual imperium.[3]

The new era metamorphosed the old one. God, gold, and glory were secularized and broadened into civilization, commerce, and fame. Civilization referred to the Enlightenment, with its hunger for useful knowledge and its secularized cosmology. Commerce extended to natural resources, not simply bullion and gems; to markets for Europe's industrialized wares; and to a traffic in knowledge. And fame? Nations as well as individuals competed for prestige, and a scientific Cortés did not challenge the state as marauding conquistadores did. Quests to Timbuktu, the sources of the Nile, the magnetic poles, and Antarctica replaced the search for mythical Seven Cities, Eldorado, and as-yet-undiscovered Mexicos.

Utopias relocated to the Pacific, or when that became known, to still-unvisited realms like Deception Island off the Antarctic Peninsula that seems to have inspired Jules Verne. But mostly utopias retreated to lost valleys and hidden coves in continents, and as blank spaces on the map filled and plausible places in which to locate them receded, they decamped to the past or future. In an age enamored with Progress, the perfection of humanity looked ahead to a time to come or back to a more prelapsarian Nature.

The reconceptualization of discovery brought with it a recharacterization of discoverers. The Humboldtean vision—the French *Encyclopédie* melded with German *Naturphilosophie*—attracted many men with wanderlust and the means to indulge it. But it proved especially compelling for certain artists. By now representational art, mathematical perspective, naturalism, even if embossed with Romanticism, flourished. It was an ideal art for reconnaissance and Enlightened exploration. Artists were moving toward landscape as a genre, and in the New World, the closer to untrammeled nature the better.

Expeditions craved art to illustrate their accounts and so carried artists on ship and along trails. Joseph Banks brought two artists on the HMS *Endeavour*; William Hodges on Cook's second voyage painted Romantic canvases of Tahiti and icebergs off Antarctica; Thomas Baines made the circuit of Britain's empire from Cape Colony to Australia to New Zealand; and some artists like George Catlin obsessed with recording the New World's natives and John James Audubon with cataloging America's birds and mammals, organized expeditions of their own. In America the Hudson River School went west; Asher Durand's *Kindred Spirits* became Albert Bierstadt's *Yosemite Valley*. But perhaps the artist most closely aligned with the age was Frederic Church.

Trained under Thomas Cole, America's catalyst for landscape painting, Church was inflamed by Humboldt's example and determined to paint the scenes the great explorer had described. After all, Humboldt had even dedicated a chapter of *Cosmos* to landscape art. Church organized travel to the Andes and the Amazon and produced enormous canvases full of naturalist detail and radiant with the majesty of nature—Humboldtean geography in paint. In the United States he recorded America's natural wonders from Niagara Falls to sunset over the Adirondacks, what America had in place of ancient Parthenons and Coliseums. Later he traveled to the Holy Land, Jamaica, and Greenland. Through the middle of the 19th century Frederic Church was America's most popular painter.

Data is just data; collections of plants, animals, fossils, ethnographic artifacts, and minerals, only boxes of stuff. They need an interpretive frame to acquire meaning. Geography and natural science provided one, but so did literature in the form of personal narratives, and so did museums created to house the specimens that replaced the specie that would have gone to treasuries in a previous age. And so did art. Its exploring artists brought discovery to the public and to politicians. The Grand Manner infused nature with moral drama. Frederic Church's paintings broadcast the sensibility of Humboldtean exploration beyond the corps of discovery that sprawled across the continents. Thomas Moran's paintings of Yellowstone and the Grand Canyon helped nurture them into national parks.

Exploration had bonded with feature after feature of Western civilization. Initially, those alliances made exploration culturally powerful, then they weakened the project, as the Enlightenment faded into Modernism, as sciences turned from geographic exploration as a means to advance knowledge, as artists left realism and landscape to investigate the nature of art, as writers turned inward to the psyche. The stream of consciousness carried more literary tonnage than unknown streams draining into the Amazon or Lualaba.

The encounter remained a set piece of discovery. Over and over again the success of an expedition depended on its ability to meet, negotiate with, translate, and adapt from discovered peoples. Exploration needed natives as guides, collectors, porters; commerce needed them as buyers and traders, and often laborers; missionaries needed them as souls; monarchs, as allies and subjects. The cataracts of the Congo were minor threats compared with the suspicious tribes along its banks. Grizzly bear attacks were less hazardous to Rocky Mountain trappers than surprise skirmishes with Blackfeet. The McCartney mission to Emperor Qianlong met with little more success than da Gama's to the zamorin of Calicut ("Go home! Take back your gifts!")—and eventually ended with violence to enforce wills. Hostile indigenes, local potentates, even wary officials—exploring parties had to adjust to them as much in the Second Age as in the First.[4]

The nature of encounters changed, however, and with them, the character of those doing the encountering. Anthropology had begun as a subdiscipline of theology in the First Age, then secularized in the Second to merge with natural history, and eventually to segue into social science. Just as geology and biology had been reinvented to cope with discoveries in natural history, so ethnology and ethnography emerged to understand newfound tribes—the terms were coined in 1828 and 1834, respectively. Archaeology, the material study of people in the past, joined them in 1837. Newly encountered peoples were viewed through the revelations of a True Science rather than a True Faith. New peoples existed as curiosities in their own right, to be mapped, studied, arranged into larger panoramas of humanity's place on the planet. They were inevitable features of explored lands, as much as mountains, birds, and flowers, inescapable and necessary parts of discovery. Then they became objects of discovery in their own right. Learned societies devoted to their study arose. It's hard to imagine those disciplines developing in the absence of a constant stream of newly found societies.

Discovered peoples were no longer classified solely as believers or nonbelievers or even might-become-believers but according to hierarchies mostly determined by their technological achievements and the sophistication of their art and literature, as Europe judged them. Europeans claimed the highest rungs; Fuegians and Australian Aborigines were close to the bottom. Newly discovered folk were fitted into the great chain, racial traits became shorthand for rankings. This was also an age that discovered deep time; over the course of the Second Age, the known longevity of the Earth increased a millionfold. Inevitably, the categories of nature were similarly historicized. The pursuit of new peoples became a search for missing links in evolution's chain of being.

That project extended back into history. As explorers crossed deserts and hacked through jungles and dug amid ruins, they discovered peoples who had once flourished and were now gone; and as they studied rocks, they found lost species, biotas, and human ancestors now vanished. Explorers trekked across blank time, not just over blank space. In his 1735 edition of *Systema Naturae* Linnaeus listed four variants of *Homo*, divided by skin color—Europeans (white), Americans (red), Africans (brown and black). In the famous 1758 tenth edition he added wild (*ferus*) and monstrous, including troglodytes. Those cave-dwelling creatures would become Neanderthals, erectines, habilines, and other sires and cousins to *Homo sapiens*. The sapiens enjoyed an ever-exploding number of entries as culture supplanted genetics; language replaced fossil jawbones; and the relics of past achievements, the oft-melancholy chronicle of people who had come and gone. The Second Age was as fascinated by the monuments of Easter Island as by the Pyramids, the ruins at Chichen Itza and Ek Balam as those of Persepolis and Pompey. The last of a breed—whether Mohican, Tasmanian, Amazonian, or passenger pigeon—became a literary convention, and the search for it, a quest of discovery.

But people were more than fossils or granites. Encounters were moral moments: they demanded judgments about how people, the discoverers no less than the discovered, saw themselves in the great scheme of things. Europeans tried to impress other societies, and then reform those peoples, through Europe's science and technology. Steel, steam, telegraphy, Maxim guns—these were the cherished emblems of an industrial society, the motive powers behind European expansion, and the means to baptize benighted peoples into modernity for their own good, whether they wished it or not. Secularization, too, had its forced conversions.

Yet encounters could challenge those assumptions. There were peoples who did not seem to divide into fallen and redeemed, who had different, often more relaxed, arrangements for living, who seemed closer to nature. The hedonistic Tahitian replaced the Brazilian cannibal as a preferred symbol for contrasting European culture with others. And if the modern sciences seemed to justify imperialism, as religion had in an earlier era, they also came to protest against it, as again religion had. By the 20th century the notion of cultural relativism had matured and by midcentury triumphed after Europe had ceased to project its racial profiling outward onto colonies and turned it inward, back on itself, most notably with Nazi Germany. By then imperialism had reached its zenith and was set to implode.

Ethnography became not merely an inevitable byproduct of exploring, but anthropologists began sponsoring expeditions on their own to search for the overlooked, if not the lost. Often the early forays fossicked for ruins and relics of peoples from the distant past—to classic sites of ancient history like Troy, or to relics mentioned in Scripture but not translated from text to stone, or New World equivalents like the vanished peoples who built cliff dwellings in the Southwest and mounds in the Ohio Valley. But there were apparently lost tribes as well, and people who had been encountered but not really studied like the Andaman Islanders, the Inuit, and the innumerable tribes tucked away in the deeply crenulated terrain of New Guinea.

It is no accident that anthropology's founders were typically men from nations with a vigorous tradition of exploration. In *The Golden Bough* (1890) James Frazer managed to combine antiquity with modern discoveries. A founder of cultural relativity, Franz Boas, inspired by Humboldt, began his anthropological career with immersive expeditions to the Inuit, before enlarging that realm to include the Kwakiutl and other peoples of northwestern North America, and finally by establishing a school of anthropology, complete with society and journal, that dispatched students to map and mop up the still-unreported or inadequately recorded tribes of humanity. When the supply of peoples dried up, as it did at the poles (and vanished outright in Antarctica), anthropology had to turn to the myriad "tribes," including occupational, that populated modern society. By then it had morphed into scholarship shorn from exploration.

Each great age of discovery had its characteristic response. Ferdinand Magellan tried faith healing and mass baptism in the Philippines. James Cook and Joseph Banks collected artifacts and tried to place Polynesians into natural and human history. Bronislaw Malinowski and Margaret Mead lived among and studied Trobriand Islanders and Samoans, respectively, for academic theses. When exploration revived by traversing Antarctica, the solar system, and the deep oceans, ethnologists had to examine the explorers themselves and, in the case of Mars, their interaction with robots. As explored lands emptied of to-be-discovered tribes, encounters turned inward.

The First Age had been charged with finding routes to the Indies. It did that, and more, and completed the task by the early 18th century. The Second Age was charged with inventorying the Earth through Enlightenment eyes. By the early 20th century it had accomplished that goal. By the time it wound down only fragments of the Antarctic coast remained unmapped along with most of Antarctica's interior, sampled only by a few traverses, but that seemed enough. East Antarctica was an immense ice sheet the size of Australia—there was not much there to know. The ice buried even continent-spanning mountain ranges.

When the Great Voyages started, every people on the planet had its own map of the world. When the Second Age ended, Europe's map was the world's. Exploration fed on itself, or to borrow an image from the machinery revolutionizing industry, it resembled a self-reinforcing dynamo. Extended expeditions carried libraries with them, stocked with books of previous explorers. On the eve of the Great War, with European imperialism at flood tide, the West could point to its undaunted tradition of discovery as a reason for its rise. Rule followed trade, trade followed the flag, the flag followed the West's intrepid explorers.

The Second Age was as varied and capacious as Europe's ambitions. The hardships and triumphs of its explorers became the stuff of modern legends. But in the end what made the age distinctive was the mind which drove and absorbed those treks and voyages. Looking back on what he considered "the worst journey in the world," with one ordeal tumbling into another—Robert Scott's *Terra Nova* expedition to Antarctica—Apsley Cherry-Garrard thought that the vital core was an "Intellectual Passion." True exploration had to go beyond physical adventuring: it had to engage the mind as well. And behind both was a moral urgency that discovery mattered and was worth the costs, the suffering, and the ethical angst it prompted. It had to interbraid with the larger culture. So, to commemorate the death of Scott's polar party, Cherry-Garrard proposed to raise a wooden cross at Observation Hill on McMurdo Island, part of a continuous lineage of memorial markers that traces back to the Portuguese padrão. He chose for an inscription the closing lines to Tennyson's great paean to exploration, "Ulysses": "to strive, to seek, to find, and not to yield."

That seemed a suitable testimony at the time for an expedition whose drive and ambition had taken its members to the ends of the Earth, to places where geology was reduced to a single mineral, where biology and society (other than the exploring party itself) was banished, where the seasons simplified into a single long day, where there was little to gather, little to see, little to learn, where the ice reflected back on the looker, where there was plenty of striving but not much to seek and less to find. Yet in its attempt, *Terra Nova* went not only to the ends of the Earth, but to the end of the Second Age. There it had, slowly, fitfully, grudgingly, to yield.

5

Alexander von Humboldt
Ascends the Heights

I have the crazy notion to depict in a single work the entire material universe, all that we know of the phenomena of heaven and earth, from the nebulae of stars to the geography of mosses and granite rocks—and in a vivid style that will stimulate and elicit feeling. Every great and important idea in my writing should here be registered side by side with facts.[1]

—ALEXANDER VON HUMBOLDT, ON HIS AMBITIONS FOR *COSMOS*

As with other great ages, the Second had its grand gesture. But so varied was exploration, so complex was its interweaving with the larger culture, so dense its mass, that there are many points of departure, sites of synthesis, or defining gestures possible. It can be difficult to identify a single style as characterizing the era. At which of these points does synecdoche become symbol?

Yet there is one ambition, one style of exploration, which seems to transcend and define the rest and which the age regarded as its highest expression, and that is the traverse of a continent. Such a project summarizes in map and narrative the useful, the marvelous, the abundance, and the hazards of lands that were either unknown to scholarship or simply unknown altogether. Each continent had its saga, some expeditions that failed and some that succeeded, but until that crossing occurred the raw work of reconnaissance remained undone. The continental traverse, enlightened by the apparatus of science, was to the Second Age what a circumnavigation of the world ocean was to the First.

But why not the ocean? Why not consider the resurvey of the world's seas the beacon for the age? Why not point to those fabulous circumnavigations done under the prod of French-British rivalry that sailed with science on their quarterdecks? Why not consider the HMS *Beagle*, reducing Magellan's Strait to cartographic rigor, carrying Charles Darwin on his epiphanous voyage 'round the world, redefining the significance of islands at the Galápagos? Why not point to the *Challenger* expedition, a remake of the *Victoria* but with scientifics rather than conquistadores? Why not consider the biography of a voyager like Captain James Cook rather than single out a style of expedition? Why not, indeed?

The oceans mattered—were themselves objects of interest. But there are good reasons to focus on the continents—that while ships often took expeditions to their overland departures, the voyage itself was secondary; that an expedition could be replicated in ways the life of an individual explorer could not; that the dynamics of renewed European expansion were not simply exercises in imperialism reenergized by the industrial revolution but expressions of settler societies that offered not only places for discovery but a more salutary moral drama. There was far more to be learned from land than from sea. The continental traverse became, for the era itself, the grand gesture of its ambitions, style, and scope as its spread from Europe, continent by continent, to the poles.

Behind them all loomed the towering presence of Alexander von Humboldt.

The Second Age passes through him like a cipher. Born into minor Prussian nobility, he had an exemplary Enlightenment education through tutors, then joined his brother Wilhelm at Göttingen. There he met Georg Forster, still aglow from Cook's second voyage; together they traveled around Europe, visiting Paris during its revolutionary fervor and meeting Louis de Bougainville, returned from his circumnavigations, and then on to Britain, where they met Joseph Banks. The journey along the Rhine led to a scientific treatise. His meeting with Banks led to a lifelong friendship. Humboldt then furthered his preparations by learning languages at Hamburg; anatomy, astronomy, and instruments at Jena; and geology at Freiburg, where he studied under A. G. Werner and met Leopold von Buch. After graduating he worked as a mine inspector, came to the attention of Goethe, with whom he also became friendly, and joined the cadre of Weimar Classicists. He journeyed through Switzerland and Italy, inventorying rocks and plants. Seeking a foreign voyage, he circled back to Paris. He was scheduled to join Nicolas Baudin's circumnavigation, but that fell through, and with Aimé Bonpland he sought to enlist in Napoleon's Institute of Egypt, but that too failed when France denied permission. Like Columbus and Magellan before him, he turned to Spain. Revealingly, during his pilgrimage to the court of Carlos IV, he measured and mapped, and calculated, for the first time, the altitude of Madrid. His timing was good—the short-lived Spanish Enlightenment and Bourbon reforms were in bloom, with the monarchy sponsoring a series of scientific expeditions to New Spain and South America and the Malaspina Expedition around the world. Humboldt's petition came without political overtones, since he would travel as a private citizen, and without a request for expenses, since he would finance the trip.

He went no place truly unknown, but such was the power of his measurements, collections, and insights that he revealed those places as if for the first time. At the Canaries he hiked up Pico de Teide and went beyond collecting curiosities like the dragon tree, and arranged species systematically in what later was termed life zones, a project that made Teide into an index for doing natural history and a reference point for organizing the Earth's mountains. In Ecuador he examined equivalent volcanoes, some still active (like Cotopaxi), others towering over the land like obelisks. He mapped Mount Chimborazo in more elaborate means than those he tested at Teide and translated its slopes into a cartography of natural history. In the process he made Chimborazo as much a symbol of the Second Age as Teide had been for the First. As he had done with Madrid, so he now measured Mexico and other regions of Spanish America with Enlightenment metrics. He remapped familiar routes not only with latitude, longitude, and elevation but with new principles of geographic organization along lines of temperature, pressure, and magnetism. He collected specimens at an almost industrial level to fill out the great chain of being, to which he gave a second axis to make a grid. He experimented with electric eels. He sought out a waterway that reputedly connected the Amazon and Orinoco watersheds—militarily and economically worthless, but a gem for geography. He visited Cuba, hiked across Mexico, and then trekked where his curiosity beckoned. Everywhere he was lionized. He was one of the most painted figures of his generation. What Napoleon was to the politics of the era, what Beethoven was to its music, Humboldt was to science and exploration.

If the French *Encyclopédie* was one end of his surveyor's rod, German *Naturphilosophie* was the other. His scientific output was prodigious: he was a one-man Institute of Egypt. His 51-volume *Voyage to the Equinoctial Regions of the New World* paralleled the *Description of Egypt* as a model of comprehensive inventorying and validated the New World as a place with its own pyramids and ancient civilizations and endowed with a nature far more astounding than the Nile-nurtured desert. His *Political Essay on the Kingdom of New Spain* advised the Bourbon monarchy about what, at the intersection of physical geography and moral philosophy, it ruled and how to more rationally administer it. His *Personal Narrative* relocated the sea travelogue onto land. His Thomistic *Cosmos*, conceived in 1834, aspired to summarize the knowledge of nature that Enlightenment science had unveiled. Later, he traveled to Siberia at the request of Alexander II and did in weeks what he had taken years to do in the Americas. Ralph Waldo Emerson proclaimed him a modern Aristotle. Throughout he pondered what he regarded as the "great problem of the physical description of the planet": to determine "the laws that relate the phenomena of life with inanimate nature."[2]

Yet the Romance was there in equal portion as he traipsed up mountains, encountered monkeys and jaguars, sought out the Casiquiare canal deep in Amazonia, exulted over immense vistas, and analyzed the distribution of flowers, all of it infused with a sense of the wholeness of nature's grandeur and its power to inspire. Meticulous with details, unbounded with panoramas—Humboldt fused in his own career the great themes and enthusiasms of his time. His *Aspects of Nature* modernized the natural history essay; his *Personal Narrative* inspired a global generation of explorers, not least Charles Darwin (it was one of three books Darwin took with him on the *Beagle*); his *Views of the Cordilleras* transplanted notions of the sublime from the monuments of ancient Europe to the Americas. In his preface to *Aspects* he announced that "a far-reaching overview of Nature, proof of the cooperation of forces, and a renewal of the delight that direct experience of the tropics gives to a person of feeling are the goals to which I strive," along with revealing "the eternal influence that physical Nature exerts upon the moral disposition of Humanity and upon its fate." Here, truly, was the explorer as Romantic hero.[3]

Critically, he undertook the project in his own name and at his own expense. (He had inherited a sizeable estate, but exhausted it paying for his travels and publications.) He represented no state. He spoke for no commercial company chartered to extract resources, organize native labor, or control trade. He aimed to convert no one. He did not sequester his findings as state secrets but published them to the world. He stood for science, as purely as anyone could. A passionate democrat, he condemned slavery; an astute observer of the natural world, he denounced environmental wreckage. His power was the power of data and ideas. No army trailed behind him. No country would use his discoveries to announce claims. He became an exemplar for others, particularly those Germans who had at the time no colonial state and either financed their own travels or joined the expeditions of other countries.

Of course he was unique: few could match his private wealth or afford his passion. But he showed what explorers and exploring nations might do, and as an ardent democrat sympathetic with an age of revolution, his exemplar could be openly emulated in ways that the institutions of despots, however enlightened, could not. He stood as the ideal. He corresponded with everyone, his residence became a salon, and a visit to him was expected as part of a new grand tour of Europe. For at least the first half of the 19th century, explorers, especially American and German explorers, were, as William Goetzmann put it, "Humboldt's children."[4]

The English translator of his *Personal Narrative*, Helen Maria Williams, aptly captured the sentiments of his readers, even channeling some of his prose style. "How happy the traveler, with whom the study of Nature has not been merely the cold research of the understanding, in the explanation of her properties, or the solution of her problems! who, while he has interpreted her laws, has adored her sublimity, and followed her steps with passionate enthusiasm, amidst the solemn and stupendous scenery, those melancholy and sacred solitudes, where she speaks in a voice so well understood by the mysterious sympathy of the feeling heart." With "eager delight" Humboldt's readers could "wander in another hemisphere! to mark unknown forms of luxuriant beauty, and unknown objects of majestic greatness—to view a new earth, and even new skies!" It reads like a 19th-century prologue to *Star Trek*.[5]

A new Earth, a new heaven. For most of the nineteenth century Humboldt remained a pole star of exploration, even when discovery meant rediscovery. Since then he has become himself the object of repeated rediscovery. He has reincarnated as a comet, returning as each generation rediscovers him, with renewed awe and appreciation, as a personification of what is best about Western civilization and the quest narrative that sent it on its voyages of discovery around the globe.

6

Crossing Continents

Great joy in camp we are in *view* of the *Ocian*, this great Pacific Ocean which we been so long anxious to See.[1]

—WILLIAM CLARK, NOVEMBER 7, THURSDAY, 1805

I t's easy to identify for the Second Age the contributing constituents that combined travel with adventure, science, art, and literature; to note how narrative came to mean more than diaries, anecdotes, or rosters of data; to see how such treks could interbreed with geopolitical ambitions, whether those of new settler societies or of outright imperialism; to watch the figure of the explorer morph from religious crusader or commercial adventurer to naturalist in the service of nationalism or the secular equivalent of missionizing monks, obedient to their particular order of scholarly discipline. But those pieces had to combine in a way that spoke to the needs, values, and understanding of the sustaining culture. A cross-continental traverse did that.

A trek across Asia or Australia was an accomplishment that could bring fame. It could reveal gross geography and make tangible what travel across it might cost; but a trek could also go nowhere, and plenty of journeys did. They were stunt, adventuring and extreme sport, but also for a few a personal epiphany, and they bequeathed some of the most riveting stories of survival in all of Western literature. To share in the age, to present the features of the syndrome, however, they had to have more than personal meaning. They needed a social purpose, and ideally an intellectual one. The paragon of an explorer was not just an adventurer who journeyed wide or hard with maybe some cultural baggage, but a scholar, artist, scientist—a member of the literate class, or at least an emissary of the elite of the day—who used travel to advance a political, national, intellectual, or broadly cultural agenda.

A cross-continental traverse announced the first raw reconnaissance. Others emulated the feat by trekking in different latitudes or along meridians of longitude. Each continent posed its peculiar problems and promises. Across the Earth the era begins in the latter half of the 18th century, climaxes in the 1870s, and slopes to a denouement, though a spectacular one, in the early 20th century at the poles.

EUROPE

The exploration of Europe sounds implausible because Europe was the homeland of the age, and furthermore, it is masked because it usually gets cast with the history of science. Europe was the Old World, after all. Real explorers sought new ones.

But Carl von Linné around Sweden and Leopold von Buch in central Europe were explorers as much as James Cook and Joseph Banks. The Linnaean apostles who ventured to the outposts of Europe's commercial imperium learned their craft in Uppsala and by discovering new species and reviewing landscapes in northern Europe. The geologists who mapped rocks, traced the contours of an ice age, and climbed mountains learned the basics in Britain, France, and Germany, ascended the Alps, and carried disputes between Neptunists and Vulcanists from Freiburg to Quito. The artists who recorded Tahiti and the eruption of Cotopaxi in the Andes followed those who had painted pastorals, Alpine panoramas, and the rumblings of Etna and Vesuvius. The travelers who unveiled lost civilizations in the Yucatan, the Andes, and Easter Island were projecting from the ruins on the pilgrimage route of the Grand Tour at Rome and Greece, and comparing them to Egypt. Alexander von Humboldt not only honed his skills when he journeyed to Madrid, but made a new map with accurate elevations. Alfred Wallace readied himself for Amazonia by surveying, collecting, and sleeping in caves in Wales, whose natural history was as little recorded in formal learning as that of South America. Torneå was as intellectually remote as Tobolsk, the peasants of the Pyrenees potentially as exotic as those of Polynesia.

The Enlightenment exploration of Europe proceeded in choreographed step with the early voyages and treks that defined the age. The great sciences of natural history took modern form not just in parallel to but in symbiosis with the Second Age, as practitioners moved forcefully out of texts and into laboratories and field studies; the cabinet of curiosities evolved into national gardens and natural-history museums. The Jardin des Plantes was established in Paris in 1626 to hold medicinal species, then morphed into a research powerhouse, the Muséum National d'Histoire Naturelle, under Georges-Louis Leclerc, Comte de Buffon, a contemporary of Linnaeus. Kew Gardens resulted from merging royal estates in 1772, adding some collections by Joseph Banks, then becoming a national botanical garden in 1840, by which time species were pouring in from British collectors around the world. A rediscovered Europe became the norm by which to compare and classify lands at the ends of the Earth. The rocks that defined and named every period of the geologic time scale were almost all found in Britain; the eras were named by British geologists. The herbaria that established the great chain of being, later revised into more "natural" associations, were assembled from European sources. As students, then mature scientists, visited one center of the learning after another, they concocted a Grand Tour for scientists.[2]

The great European explorers learned their craft in Europe. But that also meant they took Europe as a reference point. This allowed the comparison and contrast so vital to understanding and at the core of the methodologies of the day. It also led both those who traveled and those who stayed in lab and study astray when Europe proved not normative but anomalous. Exploration might spark appreciation for ethnocentrism, but it did not mean explorers could overcome it, or wanted to. Enlightenment exploration secularized but did not abolish the missionary impulse.

SOUTH AMERICA

South America's exploration was also a rediscovery. Within 50 years after Columbus's first voyage, 20 years after Cortés conquered Mexico, Spanish explorers had crossed Mexico, Panama, and, summiting over the Andes, Brazil (and for that matter, North America as well, with two half-continental traverses from De Soto and Coronado, and the crazy odyssey of Cabeza de Vaca). No other continent experienced anything like it.

Then it stopped. Over the next 250 years there was infilling, primarily by missionaries and, throughout the Amazon Basin, by *bandeirante* slavers. But the great task was done: there were no more Aztec or Incan empires to fell and plunder. The routes of trade and travel were known, the political geography of indigenous peoples mapped. There was no reason for further exploration, much less by freebooting conquistadores. What Spain wanted to know about its American colonies, it knew. Not least, Spain didn't want anyone else to know what it knew. It suffered enough along its coasts from pirates; it didn't want to advertise its interior holdings.

The French Geodesic Mission was an exception, though also an exemplar. So perhaps was the journey to Baja California to measure the 1769 transit of Venus, under the direction of Joachín Velázques de León of Mexico and the Abbé Chappe d'Auteroche from the Paris Academy. Then, as the Bourbon monarchy settled in, and the Enlightenment seeped into Spain as throughout Europe, interest rose in reevaluating the resources of Spanish America and using the new sciences to promote development and strengthen administration. Carlos III agreed to sponsor a series of expeditions outfitted with the apparatus of the emerging sciences or at least a commitment to a more rationalist perspective. They were, in a sense, to Spanish America what Russia's series of imperial expeditions were to Siberia.[3]

There were three organized for the continent, all officially designated "botanical expeditions," all extending from 11 to 25 years. The Expedition to the Viceroyalty of Peru (1777–88) traversed the Andes and west coast from Peru to Chile. The Expedition to the Viceroyalty of New Granada (1783–1816) surveyed Columbia and Venezuela. The Expedition to New Spain (1787–1803) toured Mexico and overlapped with Humboldt's circuit. All contributed specimens to the Royal Botanical Garden in Madrid, and all (including Humboldt later) issued recommendations for improvements. In addition, Spain joined the circumnavigation craze with the Malaspina Expedition (1789–94) whose special political goal was to firm up Spanish claims to the Pacific Northwest of North America. By then that other offshoot of the Enlightenment, revolution, both in Europe and the colonies, was set to boil over and cost Spain her American holdings.[4]

However inspirational the Enlightened explorers, they were not supported by Spain's successors, nor did Latin America develop the density of scientific institutions and sponsors that North America did. But then it didn't need them since most of the lands and people were already known; they were just slow to be translated into the language and matrices of modern science. Further exploration proceeded by individual collectors like Henry Bates and Alfred Wallace in the Amazon Basin, Henri Coudreau and Augustin de Saint-Hilaire in Brazil, and Germans on Humboldt's example like Prince Maximilian of Wied, or by naturalists on circumnavigating expeditions from other countries, of whom Charles Darwin is the most celebrated. Instead, exploration segued seamlessly into simple field science. Revealingly, none of the postindependence explorers were from the country they explored. The "splendid isolation" that came to characterize South America's fauna might also apply to its exploration.

NORTH AMERICA

Of all the habitable continents North America enjoyed the earliest and most vigorous exploration. Reasons are not hard to find. It was not overflowing with people like much of Asia, nor laden with diseases like Africa. Unlike Australia it had river access to the interior and landscapes hospitable to Europeans. Unlike Antarctica it was not empty of everything save ice. Not least, as the First Age wound down, only India had as many settlements from as many nations; and that imperial rivalry not only continued into the Second but acquired a new player, the first of the independent settler societies, as the United States became, in William Goetzmann's phrasing, "exploration's nation."[5]

The European contest for empire first involved a search around or through North America. By the late 18th century four nations had claims—Spain, France, Britain, and Russia. The latter three countries had settlements far enough north that Spain was unable to contest them, though in the Pacific Northwest all did converge, with exploration as a medium of rivalry. Britain sent Cook's third voyage to map the western coast of Canada; Spain dispatched the Malaspina Expedition to reassert its claims; Russia established quasi-colonial presence through the commercial interests of the Russian America Company; and the United States had whalers and merchants interested in furs and the China trade prowling the shore (in 1792 Robert Gray sailed into the mouth of the Columbia River).

Mostly exploration was maritime, tracing the coasts. Worried about its interior borders, Spain renewed its classic mission-and-presidio model, sending tendrils up California and into the Southwest, and organizing surveys to find ways to link New Mexico with the coast. In Canada it was furs and missions rather than bullion that drove forays inland; the rivalry between Britain and France added political and commercial urgency, and led to two different styles, epitomized by the French North West Company (NWC) and the British Hudson's Bay Company (HBC). The British hugged the coast, letting fur-laden natives come to them in New York and around the factories that lined Hudson Bay ("asleep by the frozen sea," as one disgruntled parliamentarian put it). The French went to where the furs were. They probed down the St. Lawrence and Ottawa Rivers, around the Great Lakes, the voyageurs blitzing with canoes and portages over the Canadian Shield much like Russian promyshlenniki crossing Siberia. Between missionaries and voyageurs the French discovered Niagara Falls in 1649, unraveled the hydrography of the Great Lakes that fence the Shield, and in the 1670s encountered the Mississippi River, with LaSalle completing a voyage to its mouth in 1682. Others pushed west—north into the fringes of the boreal Shield and south into the Great Plains. When New France was absorbed by Britain after the Seven Years' War, the rivalry continued through the competing styles of the fur trade, one sited in Montreal and the other around Hudson Bay, which meant the pace of discovery persisted.

Then the American Revolution jolted the imperial dynamics. A vigorous American empire upset the balance of power throughout North America, an equilibrium that could not be resolved in Europe through European wars. Whatever other rivalries they might have, the European imperialists each in turn had to deal with the United States, a nation founded during the Enlightenment, whose politics was based on Enlightenment ideals, and whose aggressive exploration epitomized Enlightenment discovery fused with folk wanderlust.

Instead competition intensified throughout the continent. One of the North West Company's celebrated factors, Alexander Mackenzie, was the first to traverse the continent, doing it twice. In 1789 he paddled down the eponymously named Mackenzie River to the shores of the Arctic Ocean; in 1793 he crossed the Rockies and reached the tidewaters of the Pacific. Hudson's Bay Company struggled to keep up. One of its most ambitious explorers, David Thompson, eventually left it for its rival, and then completed a route to the Pacific in 1810–11. Despite its legendary prowess, the North West Company's supply line was so long that it could take two years to complete a circuit. For different reasons, but from a common competition, both companies faced ruin. In 1821 they merged into a consolidated Hudson's Bay Company. The revived monopoly then competed with American rivals throughout the Oregon country and the interior West.

In the northwest the Russian fur trade followed Bering's second voyage and Cook's third, treating the Bering Strait as little more than a big Lake Baikal, and did in Alaska what they had done across Siberia and the Far East. The institutional context was the chartered Russian America Company, headquartered in Sitka. Its agents explored Alaska's islands, its rivers, and its tangled coast. Tracking the sea otter, whose pelt was one of the few items China was eager to trade for, the Russians and their Aleut hunters followed the Pacific Coast south as far as the coastal islands off southern California; their farthest post was on the Russian River, just north of San Francisco Bay, and so conveniently beyond the reach of Spanish authorities. East of Alaska they faced difficult mountains and inland competition from Canadian fur companies. To the south they had to deal with New Spain. They negotiated boundary arrangements with both rivals and with the United States. They cultivated Hawai'i as a mid-Pacific depot. Whether or not the Russian American Company seriously contested North America, it was competing for the north Pacific. Its presence only ended in 1867 when Tsar Alexander II sold Russian America to the United States.

New Spain had left its northern frontiers to the slow progression of missions. The unrest that led to the American Revolution, however, convinced it to strengthen that now-unstable border, which it did by arranging a series of expeditions in the 1770s. There were voyages along the California coast to check Russian and British ambitions; travels by missionaries and military engineers to report on the southern plains, especially Texas; and entradas by Juan Bautista de Anza in California, Father Tomás Garcés along the Colorado River, and the remarkable Franciscans Atanasio Domínguez and Silvestre Vélez de Escalante to find a route between Santa Fe and Monterrey, California. As always Spain kept strangers out—calling oneself an explorer made no difference.

The American Revolution unsettled the relations between Britain and the United States, and the Louisiana Purchase unsettled them between New Spain and the United States. Shortly afterward two expeditions tested the dimensions of the vast but unmapped territory. Zebulon Pike journeyed southwest toward Santa Fe, only to be arrested at the putative border. Lewis and Clark (1804–6) managed to breach the northwest borders, cross the Rockies, and spend a wet winter at the mouth of the Columbia River. Here, for the young nation, was a corridor to the Pacific—its own potential "passage to India," its first transcontinental crossing, and the exemplar for future expeditions. It was followed by other traverses from the American Fur Company, and follow-up expeditions like that under Stephen Long (1819–21) to fill out the dimensions of Louisiana.

The floodgates opened in 1821 with New Spain's successful revolt against Spain. The Rocky Mountain Fur Company dispatched a party up the Missouri River to the Rockies while others pioneered the Santa Fe Trail. For the next 25 years the exploration of western North America was driven mostly by the hunger for fur: pelts were for North America what silver was for Latin America. Interestingly, Jedediah Smith, the mountain man who trekked most widely, including three western crossings (north, south, and through the Great Basin), began as a young member of the 1821 Ashley party up the Missouri and died while trying to blaze a cutoff along the Santa Fe Trail. His career lasted a dozen brief but eventful years. Revealingly, though literate, he never completed the promised map of his traverses. In that he might stand for the era overall with its extraordinary sagas and dramatic encounters and its fitful means to get that knowledge into formal culture.

The Mexican War established a new southern border for the United States and a treaty with Britain, the northern border. The United States needed to know what exactly it had acquired, and the discovery of gold in California made it imperative to establish overland connections with the Pacific Coast. It was not obvious that America would become a continental nation that stretched from sea to sea; more likely it would balkanize, splitting off or inspiring sister nations. (The concept of "Manifest Destiny" was coined to describe the prospect that the United States would expand by catalyzing a succession of new nations, much as it had Texas.) The suddenly bicoastal character of the United States and its long-standing interests in whaling, sealing, and trade with Asia encouraged the U.S. Navy to undertake extensive voyages of discovery, with the U.S. Exploring Expedition (1838–42) as the country's bid to join the circumnavigations of the great powers and so establish its credentials and status.

This time the army led, though often with fur trappers as guides. The Army Corps of Topographic Engineers was established in 1836, the year of the Texas rebellion. It was dissolved during the Mexican War, then reconstituted for what has aptly been termed the Great Reconnaissance. These were expeditions, not just military pass-throughs, but outfitted as a company of savants, to trace the new border and develop routes to the West Coast. Their most striking expression was the four expeditions of the Pacific Railway Surveys (1853–55) designed to blaze a route for a transcontinental railroad that would bind California to the rest of the Union. With the country increasingly paralyzed by issues over slavery, the choice of a route was politically charged because a southern route would favor the South and a northern route the North. It was hoped that scientific exploration would decide the issue—a quixotic ambition. In the end, Congress chose the middle path from St. Louis to San Francisco, but only in the midst of a civil war that had removed the obstructionist southerners.

The war itself terminated state-sponsored exploration on a national level. Instead it fell to California to establish its own multiyear survey, what began as still raw reconnaissance (half the Sierra Nevada peaks were first climbed during this time) and ended as a state bureau of geology. After the war one of the young veterans of the survey, Clarence King, convinced the army to sponsor a national survey on the California model along the route of the soon-to-be-completed transcontinental railroad from the Sierra Nevada to the Rocky Mountains. The Survey of the 40th Parallel quickly galvanized three others, all competitors, what became known as the Great Surveys, that completed the mapping of the interior West. The Army Corps of Engineers had a survey; the Department of the Interior had one; and much the smaller, the Geographical and Geological Survey of the Rocky Mountain Region, was nominally under the auspices of the Smithsonian Institution. In 1869 the latter, under John Wesley Powell, discovered the last still-unknown river and mountain range in the continental United States. In 1879 the four surveys were consolidated into the U.S. Geological Survey. Reconnaissance had slid into formal surveys and then into normal science. Not until gold was discovered in Alaska was there further exploration in the classic manner.

The political evolution bequeathed North America with only three countries. The American Civil War left the United States intact; the British North America Act of 1867 consolidated disparate colonies into the Canadian confederation; and Mexico, which succeeded in its bid for independence in 1821, endured a succession of civil wars from 1858 to 1861, and then resisted French intervention from 1863 to 1867. In that latter year the United States acquired Alaska by purchase from Russia. With the fixing of stable borders exploration wound down.

Its exploring heritage survived. The United States had begun as a discovery, had matured through further discovery, and has remained a keen proponent of discovery. Exploration's nation, indeed.

AUSTRALIA

The last of the habitable continents to be discovered, much of whose lands were inimical to European agriculture and settlement, Australia posed special challenges to colonization and unique opportunities for exploration. For much of Australia, across most of its history, explorers and their allies, prospectors and graziers, were the only European presence. In perhaps no other country do explorers loom so large in the national memory.

The Dutch discovered its western coast and then its southern while probing for routes to the Indies. Luís Vaez de Torres sailed the strait between Australia and New Guinea in 1606–7, and 25 years later Abel Tasman found Van Diemen's Land (Tasmania) and then New Zealand before looping north and west to Batavia. That put some absolute bounds on the southern continent, but its coasts were far from mapped. That task launched with James Cook's first voyage, in which he traced the eastern coast (and circumnavigated New Zealand). Based on the recommendations of Joseph Banks, Britain planted a penal colony at Botany Bay in 1788. Matthew Flinders completed a circumnavigation of Australia in 1801–3. Other penal colonies followed; and then free settlements at Swan River (1829) and Adelaide (1836). All argued against dispersed discovery. Neither colonization by convicts nor Wakefieldian close settlement—English villages transported to the Antipodes—encouraged either dispersion or its organized expression, exploration.

What changed the calculus was economics; first sheep, then gold. In 1805 Merino sheep were introduced. Unlike cattle which had to be slaughtered to get meat, hide, and tallow, sheep could be sheared, which allowed flocks to multiply, which pushed shepherds to search for new grazing grounds, even over the Great Dividing Range. The quest for pasture pushed and pulled graziers around the continent. In 1851 gold was discovered and diggers, like drovers, fossicked over any place they could reach. The presence of wandering hordes prompted government to dispatch explorers and surveyors to the new lands to better understand what there was and to impose some form of administrative order. By the end of the 19th century Australia held six separate British colonies and one territory. In 1901 they were merged, on a roughly Canadian model, into Australia.

Often compared to America as sister settler societies, their differences were as pronounced as their similarities. America was relatively easy to reach from Europe; Australia, difficult. Unlike America, Australia never rebelled against Britain, and looked to government for aid in a harsh land; its free-ranging squatters, miners, and pastoralists were less refractory than American counterparts, and the state established order over mining districts, which never dissolved into the vigilantism that characterized early California. Until the 20th century, Australia's separate colonies remained independent, so there was some competition between them but less urgency to find ways to bind them through waterways (which didn't exist) or railroads (prohibitively expensive). Nor was there a frontier that rolled or leapfrogged across the continent. Australia was coastal and urban. Despite its delight in its bushman image, Australia was populated overwhelmingly by urbanites, and the colonies were mostly city-states with large hinterlands. In each colony the movement tended to be from the coastal city to the interior, and then back again. Early exploration resembled polar discovery, with a trek to the center and back. Much of the interior was as much a hot stony desert as the poles were a cold icy one.

Its explorers were both necessity and exception. By midcentury they were eager to cross the continent. A rivalry developed between South Australia, which sponsored John McDouall Stuart, and Victoria, which assembled a massive exercise, the Victorian Exploring Expedition (1860–61). Each determined to trek north from their respective capitals to the Gulf of Carpentaria and back. Stuart had been to the interior before, and had served with Charles Sturt's explorations; his party had several false starts before finally completing the task in 1861. Meanwhile, the Victorian expedition, best known for its leaders, Robert Burke and William Wills, promised to be Australia's answer to Humboldt and Lewis and Clark. It succeeded in reaching the gulf, barely, and then ended in disaster, with but a single survivor, John King. Their failure to return on schedule inspired rescue operations, and after the revelation of their deaths, stirred a collective frenzy to complete the task. Between them their discoveries blazed droving routes and a path for a transcontinental telegraph to Darwin. That left the traverse, east and west. Edward Eyre had journeyed along the south coast in 1840–41. The interior crossings waited until the 1870s when John Forrest, Peter Warburton, and Ernest Giles (twice) made the trek between the west coast and Stuart's track.

There remained lots of geographic mop-up, but first-order reconnaissance faded. Australia's Aboriginal peoples rather than its pastures and ores became the object of interest. That observation illustrates one of the critical contributions of Australian exploration within the global context. Everywhere else, when Europeans encountered foraging and hunting societies, contact had come early by observers not interested in formal ethnography (which didn't exist as a discipline), only in souls, allies, and laborers. By the time scholars trained or interested in anthropology arrived, indigenous societies had usually been changed by trade, introduced disease, and forced relocation, if not outright war and population collapse. Anthropologist-explorers were recording societies not as they were at first contact but as they had adapted to the often brusque changes that spilled out from encounters, even secondhand contact through goods and epidemics spread along trade routes. Many cultures were agricultural, and some of course were more advanced as civilizations than Europe.

Australia was different: here Enlightenment and Aborigine met face to face. Missionaries and the chroniclers of conquistadores had recorded societies according to parameters relevant to their needs. Naturalists and early ethnographers like Joseph Banks tallied different traits. Eventually, the search for new tribes became as much an incentive for exploration as well-watered grasslands or deposits of gold and opal. Still, there was much less transfer of indigenous knowledge to Western literature than typical. More than elsewhere explorers had to see and record for themselves.

For Australia exploration mattered in ways not true for most continents. As a settler society, Australia looked to its explorers for a suitable creation story, an alternative to origins as a penal colony, and for guidance in understanding how to live on a landscape harsh and hostile by European standards. That Captain Cook was, by the standards of his time, able to muster rapport with indigenes and did not slash through the countryside with sword in hand made him a more tolerable founding father then Columbus, DeSoto, or Pizarro. Few countries have honored their explorers more roundly. And when the opportunity arose, Australia relied on that heritage to carry exploration to its own colonies in New Guinea and especially Antarctica where the Red Centre met a White Darkness.

AFRICA

There were two Africas, one north of the Sahara and the other south of it, and two eras of African exploration, during the opening fanfare of the First Age and during the imperial scramble that helped end the Second. The first planted trading factories along the coast; the second, plunged into and claimed the hinterlands. Paradoxically, Africa was the first continent outside Europe to have voyaging Europeans explore its coast and the last to have its interior unveiled. Its proximity made Africa accessible; its rumored riches and its fabled landmarks like Timbuktu, Prester John, and the Mountains of the Moon made it inviting; and its formidable environment made it remote.

Assets proved sparse, and access difficult. There were few good harbors, the rivers were full of cataracts, and veered in sometimes freakish ways—the Niger and the Congo both elbowed or arched sharply in their passage to the Atlantic. North America had the Mississippi River system, South America the Amazon, Eurasia navigable rivers (sometimes by ice), but only Australia among the other continents had rivers that proved so sparse and treacherous to navigate. The indigenes were often hostile—the Muslims of the Sahara and Sahel denied entry to Europeans, the tribes of equatorial and southern Africa were traumatized by centuries of slavers and the plagues like smallpox that they introduced and so were wary of strangers. Diseases were abundant, debilitating, and often lethal: it was impossible to travel long without succumbing to one fever or another or to several. The threat even extended to livestock: if boats were denied, so were horses, oxen, and burros, all of them susceptible to rinderpest and other ailments. The only reliable beasts of burden were human porters, so prominent a feature of expeditions and on whose good relations an explorer depended. Not least, the rumored wealth funneled into gold, ivory, and slaves, all of which could be had by trading along the coast. After the Atlantic slave trade was shut down, the avarice that drove so many early explorers waned; there wasn't enough to justify costly, onerous, and sickness-plagued expeditions.

By the late 19th century what replaced bullion as an attraction was colonies. What drove explorers was the fame that could accrue from the pursuit of legendary sites, especially the dual sources of the White and Blue Niles. What drove national efforts was the notorious partition that made Africa the scene for intense competition. Still, there were European coastal settlements at Cape Colony, Angola and Mozambique, and around West Africa that could provide points of departure into the interior.

The Sahara, closest to Europe, was the first to be traversed. The African Society dispatched Mungo Park to puzzle out the course of the Niger River. Then France and Britain competed across the West African Sahel. East Africa mostly involved working out the hydrography of the Nile and the chain of Great Lakes that led to or fed it, and because its course was so confusing, the headwaters of the Congo River. Southern Africa fell within the sphere of the Portuguese and the British. For centuries Portuguese merchants (*pombeiros*, the African counterpart to Brazil's bandeirantes) had journeyed into the interior in search of slaves; later they turned to other merchandise. One, Antonio da Silva Porto, crossed from Benguela to Cape Delgado; Hermenegildo Capello and Roberto Ivens, from Luanda to Victoria Falls, and then to Natal. Otherwise, Britons did the trekking—most famously, David Livingstone and V. L. Cameron. That left the equatorial interior—ambiguous, formidable, in turmoil from smallpox and Arab-Swahili slavers from Zanzibar. Even the course of the immense Congo River was unknown.

That changed in the mid-1870s. Leopold II assembled the Brussels Geographical Conference in 1876 to pool what was known and propose an agenda for further discoveries. But the most spectacular—perhaps the defining—journey was already underway from Henry Stanley, who proceeded to dominate the field until the 1890s. Stanley had been introduced to Africa as a war correspondent for the British Ethiopian Expeditionary force, and then achieved notoriety for his 1871–72 "discovery" of Livingstone, who became an inspiration for his further exploration in the region and for his determination to connect it with the developed world in ways that would bring—in Livingston's words—Christianity, commerce, and civilization. From 1874 to 1877, over the course of 999 days, Stanley traveled from Zanzibar to the mouth of the Congo—one of the most remarkable feats of exploration in the Second Age. A decade later he returned for the nominal rescue of Emin Pasha, which found him again traversing the continent, and had him three times crossing the fetid Ituri Forest. But Leopold's vicious administration of the Congo Free State as a personal fiefdom retarded further work and certainly set back European-African relations.

Meanwhile, the Berlin Conference (1884–85) sought to constrain colonial competition within Africa as treaties of the First Age had tried to limit quarrels between the major powers. Instead of calming political competition, however, the partition of Africa inflamed it. The European imperium was nearing its climax; only Ethiopia and Lesotho escaped colonial status, though Italy later invaded Ethiopia and the confederation of separate colonies into the Union of South Africa isolated and neutered Lesotho into a standing something like the nominally independent rajas of India. With partition came a need for boundary surveys, and this spurred a final round of discovery. By now reconnaissance—certainly the kind of epic exploration associated with Henry Stanley—was over. The scores of survey parties were laying down borders and doing field science, filling in the cockeyed web of boundaries drawn on tables in Berlin by people who had no practical knowledge of what those lines meant on the African landscape.

Among the inhabited places encountered by an expanding Europe, Africa began early and ended late. Its coast was the formative arena for testing ships, fashioning institutions, nurturing rivalries, and defining purposes; its interior was the most impenetrable of habitable lands, rivaled only by New Guinea and Tibet; the moral qualms it raised were the most strenuous since Iberians had transferred their crusading spirit to lands beyond their historical horizon. Yet even Africa followed the general trajectory of the Second Age: its grand traverses were completed by the end of the 1870s while it steadily hedged into the more routine tasks of disciplinary science. By the onset of the 20th century the blank places of Africa were being rapidly filled in by biologists, geologists, geographers, and anthropologists.

ASIA

Asia was another anomaly. It was after all joined to Europe along the Urals: together they made one colossal landmass. What interested Europe were the subcontinents and archipelagos that hung like tectonic pendants from the central Asian chain of mountains and plateaus—Cathay, India, Indochina, the Moluccas. There had been overland transportation possible since ancient times, sharpened during the Mongol imperium, then shattered when that empire of roads and bridges collapsed. Monks and merchants had trekked across the vast expanses. There had also been maritime routes, broken by seas and long portages (such as across Suez). By the 15th century they were collectively known as the Indies, were the object of European questing, and were eventually connected by vessels, ports, and trading factories.

The exploration of continental Asia proceeded in two fashions. North of the trans-Asian mountains discovery belonged to Russia. Its empire pushed eastward, led by Cossacks and promyshlenniki, punching through the Urals in 1581, then coursing rivers to the Pacific, which they reached at various points between 1649 and 1679. This was classic First Age exploration, done with pirogues and rafts rather than caravels. The Second Age geared up when

Peter the Great ordered Bering's Kamchatka expeditions to see if Eurasia and America were joined and his successors sponsored the Great Northern Expedition to map the shore of the Arctic. Bering's traverse across Eurasia was itself the first continental crossing at least nominally equipped with natural history interests since it contained academicians, notably, the naturalist Georg Wilhem Steller.

Additional treks coincided with further waves of expansion and consolidation—Enlightened despots like Catherine looked to the new scholarship to help inventory their domains and strengthen their rule. A year after she appointed Peter Pallas, a Prussian naturalist, to the St. Petersburg Academy of Sciences, she sent him on an expedition to Siberia that lasted from 1768 to 1774 and resulted in three volumes of reports. Combined with the research of others, he published summaries of Russian flora and fauna. In 1793–94 he traveled to southern Russia, including the Caucasus and Crimea. In 1829 Humboldt himself ventured into central Siberia as far as the Altai Mountains. As Britain consolidated its grip over India, and twice invaded Afghanistan, the need to protect its holdings, and a growing rivalry with Russia—what became famous as the Great Game—sparked a flurry of exploration throughout central Asia.

By 1870 the geography yet missing was the central Asian mountains and Tibet. This became the primary ambition of the Russian military officer-cum-explorer, Nikolai Przhevalsky, who "dreamed" of reaching Lhasa and who imagined himself as a Russian conquistador. "Here," he wrote of central Asia, "you can penetrate anywhere, only not with the Gospels under your arm, but with money in your pocket, a carbine in one hand and a whip in the other. . . . Here the exploits of Cortez can still be repeated." Przhevalsky traveled widely in the region and made two bold bids to reach Lhasa. In 1870–73 he traversed Mongolia to Beijing, then to the Tibetan Plateau. He tried again in 1879–80, but was turned back by Tibetans. He was readying for another attempt in 1888 when he died (possibly by poisoning). Other adventurers filled in the deserts and mountain flanks along the Russian border. Eventually, the British reached Lhasa from India.[6]

The remainder of Asian discovery belonged with the British in India, the French in Indochina (highlighted by the Mekong Expedition in 1866–67), and a miscellany in China. In truth, the whole concept of exploration in

China, outside its inner Asian claims, was improbable. Apart from the ancient silk road, by the 17th century Jesuits were already serving as court astronomers and were merging Chinese and European cartographies. The more complex story is the Malay Archipelago, whose largest isles, Borneo and New Guinea, acted as microcontinents.

Of particular value was the Moluccas, which had been a special object of the Great Voyages and whose rediscovery by naturalists illustrates the transmigration of exploration from the First Age to the Second. The story pivots on Alfred Russel Wallace, who arrived at Singapore in 1854 and spent eight years touring the isles, out of which experience he conceived the ideas behind biogeography and evolution. In 1869 he published *The Malay Archipelago*. He was on Ternate itself when, down with ague, he wrote out the theory of evolution by natural selection and sent it to Darwin (had he sent it elsewhere we would identify Wallacism, not Darwinism, with evolution). The redefined wealth of the Moluccas was not its spices but its immensely diverse natural history, the collection of which financed Wallace's travel. The critical dividing line through the archipelago was not that laid down by the Treaty of Zaragoza between Spain and Portugal but that which Wallace traced along the Lombok Strait that segregated the fauna of Asia from that of the isles and Australia to the east, later known as the "Wallace line."

WORLD OCEAN

The new perspective prompted a resurvey of the setting that had been most fundamental to the First Age. The geographical techniques that were being applied to continent after continent, even those presumed known, were extended to the world ocean. Isolated coastal charts, harbor soundings, winds and currents, and of course the shorelines of continents and islands were all consolidated and summarized. Seas that had been sailed for millennia were reexamined. Routes that had carried Europeans around the world were reexplored, this time with the apparatus of the evolving sciences.

Exploring parties mapped routes and coasts of commercial and military interest. The development of a relatively simple method for determining longitude meant that old coordinates could be updated. By the time of Cook's third voyage the coastlines of the world, save for the Arctic and Antarctic, were known; by the mid-19th century the world's islands were also identified. The U.S. Exploring Expedition sought, among various duties, to correctly map islands of interest to American traders, whalers, and sealers. (Some of its charts of Pacific islands were still in use during World War II.) The coastline was as much a map of the ocean as of the lands it washed against. State bureaus like the British Admiralty and the American navy absorbed a continuous stream of data. Mapping the waters might seem like mapping the air, yet beneath the ocean lay the largest extent of the Earth's solid surface. Dredges and soundings were devised to sample the geology and biology of the deep sea.

Interest rose in the 1850s. In parallel with army expeditions to trace a transcontinental railway, America sponsored naval expeditions to study the seas vital to its commerce. The North Pacific Exploring Expedition was, in words of Helen Rozwadowski, "the first major effort by any nation to claim the ocean scientifically." From 1853 to 1856 the expedition's naturalists redis-covered coasts but equipped with a special sounding device for sampling the bottom, it also dredged up mud and marine organisms from the depths. The United States was aggressively pursuing its two frontiers: North America and the oceans that bordered it.[7]

Appropriately, in 1855 the American naval officer and head of its Depot of Charts and Instruments, Matthew Fontaine Maury, published *The Physical Geography of the Sea*, a Humboldtean synthesis that did for the ocean what others were doing for lands. Winds and currents were plotted akin to iso-therms and lines of magnetic declination. Later editions expanded to include the oceans' meteorology, the interaction of wind and water that defined the dynamic geomorphology of the ocean's surface. But Maury was also inter-ested in the shape of its bottom, and included a rude map that plotted the bathymetry of the Atlantic.

What galvanized commercial and competitive interest was the growth of telegraphy. An undersea cable had been run from Britain to France, and from Nova Scotia to Newfoundland, and in 1857–58 a cable was extended from Newfoundland to Britain. It worked, then failed, then was relaid in 1865–66 by the fabled steamship *Great Eastern*. To span across the ocean's bottom with an expensive cable at a time when the surface of the Moon was better mapped was a daunting ambition. Still, it could stir the same kind of

Romantic enthusiasm as expeditions on land. When he wrote *20,000 Leagues Under the Sea* (1869–70), Jules Verne had a copy of Maury's *Geography* at his desk and the experience of traveling on the *Great Eastern*.[8]

The expeditions, those directly aimed at oceanography and those for which it was an ancillary task, reached a climax with the *Challenger* expedition of 1872–76. The Royal Society of London promoted the idea, the Royal Navy donated a ship. The expedition had a full complement of scientists and an acclaimed Arctic explorer, George Nares, for captain. It sailed 68,890 nautical miles, documented the physical characteristics of the world ocean, sampled its bottom, and collected species—in short, did what a century of Second Age explorers had been doing but with the onerous requirement that they do it in the depths. They measured, trawled, and dredged, over and again. Excitement yielded to tedium ("dredging" became known as "drudging"). The final report, released 1895, ran to over 29,500 pages. Thousands of specimens went to labs and museums. The ocean held more than people imagined, its contours were more complex than anticipated. The most spectacular finding, however, was an enormous canyon (8,184 meters down) in the southwest Pacific, what was named the *Challenger* Deep, and later the Mariana Trench, which plunges below sea level more than Mt. Everest rises above it. The classic circumnavigation had become a traverse of the world ocean.[9]

The *Challenger* sailed a century after Cook's celebrated voyages. Yet for all the awe the *Challenger* expedition inspired as an idea, it failed to capture public attention. It looked more like complex science than exploring reconnaissance. It met no peoples, made no first contacts, suffered more through boredom than physical danger. Verne had populated *20,000 Leagues* with monsters and villains and conflicts that turned the plot. *Challenger* experienced none of that. Its biggest problem at sea was to keep sailors from deserting (a quarter did), and at home, to rally interest sufficiently to keep publication on track for the next 20 years. It had no narrative. Unlike Cook's voyage there was no global transit of Venus to track, no lush Tahiti to rhapsodize over, no fascinating natives to interact with, no icy continent to tantalize across ice blink and fog.

As with so much of the Second Age, by the 1870s the great traverses had been completed. Oceanography progressed, ocean exploration receded. Steam-powered ships stimulated a revolution in oceanic trade and made access even to Antarctica possible. Yet the seas remained something to pass over, not explore through.

THE POLES

That left the polar regions. Arctic and Antarctic were truly poles apart. The Arctic is an ocean surrounded by continents, the Antarctic a continent surrounded by oceans. The Arctic Ocean, plated with sea ice, swirls in a slow gyre; the Antarctic is actually a continent and a cluster of islands (think Australia and New Guinea) joined by ice sheets, with an ocean that seasonally doubles the size of its ice field. The Arctic has indigenous peoples around it, the Antarctic has no one. The Arctic is accessible in ways the Antarctic never can be. The Arctic was visited in the First Age; the Antarctic, mostly in the last twilight of the Second Age. Yet the major rhythms of their exploration in the Second Age are oddly synchronized.

The Arctic has straits and isles that resembled the coarse geography that characterized what confronted the Great Voyages. A northeast passage, a northwest passage, the Bering Strait—all provided entry points for ships. But the ice was formidable and the straits only open at the end of summer; all conditions worsened by the Little Ice Age. By the mid-18th century the access points were known, and the major perimeter islands like Svalbard, Greenland, Novaya Zemlya, and the Canadian archipelago identified. Few incentives existed to entice explorers. A British Admiralty map of 1818 shows the shorelines save for northern Greenland, the Canadian mainland, and Alaska. By then the Napoleonic Wars had ended, and Britain had many ships and outstanding officers that both faced mothballing. Permanent Secretary of the Admiralty John Barrows urged that they be put to exploring; and between 1818 and 1833 a series of expeditions sailed north toward the Greenland Sea and Svalbard, and another suite to the Davis and Hudson Straits west of Greenland. One, under John Franklin, trekked across the Barren Land and northern Canada.

Then in May 1845 Franklin led two veteran polar ships, the HMS *Erebus* and *Terror* in an attempt to complete the Northwest Passage. They sent letters home with a whaling ship they met in the Davis Strait. They were never heard from again. What had been a serious but measured exploration meta-morphosed into a frenzied rush to rescue. Ship after ship launched—mostly British and Canadian, then American—to recover and then to simply find what remained of the Franklin expedition. After nine years the British Admi-ralty listed the party as dead in Her Majesty's service. Hudson's Bay Com-pany sent parties overland. The U.S. Navy dispatched ships; so did scientific organizations. Newspapers offered rewards of up to £20,000. Together they completed the mapping of the Canadian archipelago. But apart from a few relics, they failed to solve the mystery of Franklin. It was clear that the party had perished, probably dispersed across several places. The reason remained unknown, though bad ice years were a likely contributor. As an amateur quest the search for Franklin continued through the 20th century.[10]

Meanwhile, Greenland, a microcontinent, had its northern littoral explored by George Nares in the 1870s, and its southern ice sheet crossed by Fridtjof Nansen in 1888. Nils Nordenskjöld retraced the Northeast Passage across Eurasia, leaving from Norway and exiting through the Bering Strait. Nansen then tried to reach the pole by riding the icy gyre in a special, ice-adapted ship, the *Fram*. The pole was finally reached—after six attempts, among controversies that still rage today—by the American Robert Peary. Peary departed from the furthest northern land, Ellesmere Island, and used Inuit technology in the form of modified dog sleds and polar clothing. He did what most successful explorers have done—adapt native practices to satisfy European goals. He reached the pole on April 6, 1909.

Later expeditions relied on airplanes and submarines—more technolog-ical adventuring than exploring, or normal science, or the occasional reen-actment with skis and sleds. The further exploration had to wait until the ocean floor itself could be mapped.

The Antarctic had accrued more speculations than empirical visits. On his second voyage Cook had circumnavigated the continent and judged by the size of its icebergs that a large landmass, the ever elusive Terra Australis, must lie farther south. The modern age began between 1819 and 1824 but from sealers and whalers, and a Cook-styled expedition under Thaddeus

Bellingshausen that sailed for Russia. Neither Cook nor Bellingshausen managed to view the continent itself.

Exploration proceeded in two big pulses. In the first, three nations mounted classic Second-Age expeditions that converged in the Antarctic in 1837–43 and sighted the continent. Together, Charles Wilkes of the U.S. Exploring Expedition, Jules Dumont d'Urville for France, and James Clark Ross for Britain began the naming and mapping of the Antarctic coast. Then, although sealing and whaling continued, exploration turned elsewhere. Without obvious resources, without natives to serve as guides (and objects of curiosity), without a lost Franklin expedition to spur public enthusiasms and a multinational scramble for claims, the Antarctic receded into its icy silence. Then the 1896 International Geographical Congress announced that the Antarctic remained the last unexplored region on Earth and urged responsible nations to sponsor voyages of discovery.

That sparked an explosion of expeditions—the second surge—some with private and some with public financing that washed ashore in three rising waves. The first (1897–1900) arrived under British and Belgian flags. The second (1901–5) featured British, German, Swedish, Scottish, and French expeditions. The third, the fabled heroic age (1906–15), added Japan, Norway, and Australia and bequeathed some of the grandest sagas to emerge from the Second Age. Roald Amundsen reached the pole on December 14, 1911; Robert Scott, on January 17, 1912. The race to the pole; the death of Scott's polar party; Ernest Shackleton's *Nimrod* and *Endurance* expeditions; Douglas Mawson's lonely trek across treacherous ice—these provide a fitting bookend to Cook's circumnavigations, Lewis and Clark's great traverse, and Humboldt's travels among the wonders and marvels of equinoctial America.

Then exploration and empire ceased to look beyond as Western civilization was convulsed by two internecine wars and a Great Depression. The heroic age of Antarctic exploration was among the many casualties. But so was the Second Age overall. After the Second World War Europe turned inward, trying to consolidate itself rather than project its ambitions outward. Decolonization replaced imperialism. The pushes and pulls that had propelled the Second Age, more or less moribund for decades, went into reverse. Besides, so well had the Second Age done its task that it seemed there was no place left to go.

7

Second Looks, Repeat Encounters

The Grand Cañon of the Colorado is a great innovation in modern ideas of scenery, and in our conceptions of the grandeur, beauty, and power of nature.[1]

—CAPTAIN CLARENCE E. DUTTON, *TERTIARY HISTORY OF THE GRAND CAÑON DISTRICT* (1882)

Fifteenth-century Europe was far from dark, but it was much brighter after the Renaissance and the Great Voyages. When Columbus and da Gama sailed, formal learning still consisted of texts, of professors reading ("lecturing") from texts, adding glosses between the lines, all of which students wrote down. The immense project of recovering and translating lost texts was underway. It was a world of scriptures, both sacred and secular. One judged truth by comparing and weighing authorities. Europe knew parts of three continents and four seas.

By the time the First Age ended, those texts had to reconcile with real-world facts and with a novel method of creating information—reading the book of nature, which was glossed by experiment and mathematics. Geographic discovery was an ideal symbol of that change, as explorers sought out new data and then embedded it in a Cartesian system of coordinates. But modern science was still feeble—powerful in a few fields, still embryonic in most. The polymath Athanasius Kircher illustrates the kind of curious hybrid that resulted.

Kircher was a Jesuit, who taught for 40 years at the Roman College and wrote between 36 and 40 major works on nearly everything of interest to intellectuals of the time, from mathematics to languages to volcanoes, magnetism, malaria, ethics, music, and comparative religion. He was born eight years before Galileo published *Starry Messenger* and died seven years before Newton published *Principia Mathematica*. Most of the great reconnaissance of the world was done; the major works of modern science were yet to come. In brief, Kircher was the last of the Renaissance humanists, full of that age's curiosity and limited by its choice of methods to understand what caught its fancy.[2]

The outcome is, by modern standards, a mishmash, a literary equivalent to the sprawling cabinet of curiosities he maintained. It seemed he examined everything and, by modern thinking, explained nothing. He studied Vesuvius and in *Mundus Subterraneus* created a model of Earth full of internal fires that blistered the planetary crust. He discoursed on Atlantis, which he located from Egyptian and Platonic texts. In *Arca Noë* he proposed that, as creatures wandered from the ark after the Flood, they assumed new forms, perhaps as languages had after Babel. He was a Renaissance parody of Pliny the Elder.

By the end of the Second Age the First seemed steeped in superstition, if not ignorance. No one could seriously attempt to claim all knowledge as his realm. There was too much, with more pouring out each year. In time geographic discovery morphed from reconnaissance exploration to field science, as the scholar's study evolved into the scientist's lab. Cabinets of curiosities underwent a chrysalis into elaborate museums of natural history. As exploring intellectuals inventoried the land and its inhabitants, new disciplines arose in response: geography, geology, archaeology, ethnology and ethnography, and along with them international congresses and societies that replaced the far-ranging Jesuits and their Vatican college.

Consider geology. In 1783 it received a name. In 1871 the first International Geographical Congress was held; in 1878 the first International Geological Congress, appearing in syncopation with the final continental traverses, save for the poles. In 1887 the *London Atlas of Universal Geography* released its first edition, a folio of 90 maps that portrayed the "physical and political divisions" of the world. Apart from its value in systematizing the search for ore, the discipline's organizing metric, geologic time, and its informing question, the age of the Earth, had broad cultural cachet. Classic geology climaxed with the Second Age, and then entered a period of intellectual decline as the flood tide of new lands and seas, new mountains and strata, new species and biotas ebbed.

A similar scenario played out for biology (named in 1813) and anthropology. As naturalists swarmed, their collection of species filled, and demanded, museums, gardens, and herbaria. The numbers of new species led to reforms in systematics; their curious distribution, to concepts of biogeography; their changes over time, to the theory of evolution. And so it also went with anthropology. The proliferation of peoples, mores, kinship systems, languages, and myths stimulated disciplines from ethnography to folklore to modern anthropology, whose founders were travelers if not dedicated explorers. Major John Wesley Powell, after commanding a prominent survey in the American West, headed not only the U.S. Geological Survey but the Bureau of American Ethnology, the latter assembling tribes as the former did strata. To measure the distance traveled compare the contact stories of the Canaries' Guanches or Newfoundland's Beothuk to Alfred Radcliffe-Brown among the Andaman Islanders.

The great scenes of the First Age, the sites that defined the era, were places like Madeira and the Canaries, the straits of Magellan and Molucca, the South Sea viewed from a peak in Darien, the Valley of Mexico—scenes of discovery vital to a passage to India and to establishing a West Indies in a New World. The defining scenes of the Second were sites like the Galápagos Islands, Ripon Falls, the Grand Canyon, the South Pole, data-laden sites critical to new sciences, scenery valued by artists, and arenas for the sheer valor of exploring.

Additionally, the moral geographies of the Second Age held a special place for settler societies. Encounters here had many of the same ethical ambiguities as those earlier, but they were buffered and cushioned by the prospect of emerging, fresher societies that colonizing made possible. These were places not just subject to foreign rule, but to a demographic takeover by another people and culture eager for a suitable creation story. Exploration satisfied that need, since explorers could be figureheads, special symbols and ciphers. The explorer who blazed the trail for newcomers in uninhabited or lightly inhabited lands—James Cook, Meriwether Lewis, and William Clark—could be honored for making a better world possible. That most such societies really took off during the 18th and 19th centuries meant they were representatives of the Second Age with values and expectations different from those of the great voyages and grand entradas.

The discovered lands themselves might serve to validate and valorize the societies that claimed them. Not always: eastern Canada was, in Jacques Cartier's famous words, the "land God gave Cain," and was valued for tempering a tough folk. Likewise, so much of Australia's landscape was, to Western eyes, so bleak and sun-baked that the bush was mostly valued for begetting the sardonic bushman. The real expression occurred in America, which having revolted from Britain, sought its own pantheon of civilized emblems. Its Hudson River could rival the Rhine, and its Rocky Mountains, the Alps. Its predators were more robust, its landscapes much larger. Its Great Plains dwarfed the plains of Hungary, its Great Lakes mocked the Lake District, its California had a Mediterranean climate without the maladies of the Mare Nostrum. Even better, it had features like Yosemite Valley, the geyser fields of Yellowstone, giant sequoias and bison, all of which were painted in operatic canvases. With help from Thomas Moran's art, produced as part of the Geographical Survey of the Territories, Yellowstone was established as the country's first national park. Such scenes put the pastoral fields and cathedrals of Europe in the shade. Most spectacularly, America had the Grand Canyon.

It is one of the quirks of exploration history that the Grand Canyon was among the earliest of North America's natural wonders to be discovered—this in 1540 by a foray from the Coronado Expedition. A party led by Captain García López de Cárdenas trekked to the south rim. They had heard rumors of a great river, perhaps the same river that another branch of the expedition, under Hernando de Alarcón was exploring from the Sea of Cortez. They found a monstrous *barranca* whose dimensions baffled them. Three men scooted down an Indian trail and returned with the report that the buttes viewed from the rim were actually taller than the great tower of Seville. There was no passable route to the river, which was anyway unnavigable. On the rim there was no surface water and few indigenes. There was nothing more to see. The land held little of interest. It was worth remembering only as something to avoid, and so it was.

Not until New Spain sought to strengthen its frontiers in the 1770s did another European appear. Father Francisco Tomás Garcés followed the Yuman tribes up the Colorado until the river became impassable, then cut across the plateau to the Havasupai people who lived in a side-canyon, before proceeding on to Santa Fe. Later, he was killed in the Yuman uprising. No one followed his trail. The Escalante-Dominguez expedition from Santa Fe to Utah, pioneering what later became the Old Spanish Trail, assiduously avoided the greater canyon. Again, the land was formidable, it held no attractions, it was a barrier not a beacon.

American fur trappers in the Southwest knew of it, and later army expeditions looking for wagon and rail routes, often with former trappers as guides, shunned it. In 1857, however, an expedition under Lt. Joseph Ives investigated the Colorado River as a possible supply route into Utah. The party took a steamboat up the Colorado with the rumored "Big Cañon" as a sideshow. The boat sank in Black Canyon; the party moved overland to Diamond Creek, which they traced into the gorge of the western canyon, after which they followed Garcés to the south rim and on to Santa Fe. It was remarkable scenery, but in his 1861 report Ives voiced the same sentiment as those before him, that it was "altogether valueless." His, he intoned, had been "the first, and will doubtless be the last, party of whites to visit this profitless locality. It seems intended by nature that the Colorado River, along the greater portion of its lonely and majestic way, shall be forever unvisited and undisturbed."[3]

Between 1869 and 1882 that perspective was turned inside out. Between John Wesley Powell's two descents through it by boat and Clarence Dutton's meditations on the rim, what had been an indelibly alien landscape morphed into and was celebrated as a distinctively American one. The canyon seemed a mother lode for geology, lithic testimony to the transcendence of geologic time, an aesthetic marvel for art, and its landscape as exceptional as the American experiment itself. Thomas Moran painted it in a companion work to his Yellowstone masterpiece. It offered sights that challenged conventional notions—"a great innovation in our modern ideas of scenery," Dutton exulted. Had even some of its features been "planted upon the plains of central Europe it would have influenced modern art as profoundly as Fusiyama has influenced the decorative art of Japan," yet here such features are "swallowed up in the confusion of multitude." Explorers forced a redefinition of those aesthetics. Writers, painters, and nature enthusiasts then joined the chorus. Like America, a new innovation in our ideas of politics, America's Grand Canyon offered something novel and compelling. One was a fitting emblem of the other. In 1903 President Theodore Roosevelt stood on the South Rim and proclaimed it an object for national pilgrimage, declaring it "one of the great sights that every American should see."[4]

If the canyon inverted inherited expectations about what qualified as majestic scenery, so it also inverted the process by which a place came to be socially valued. The land had not been lived on for centuries, building up a loam of folklore, before intellectuals arrived and made it iconic. Rather, its value was asserted by elites and then popularized. Those elites—the geologists, the writers, the painters—encountered the scene as explorers. The physical excavation of the canyon resulted when the Colorado River, instead of running parallel to the trend of plateaus, suddenly angled 90 degrees westward and ran through them, before returning to its southerly course. That unexpected turn expresses exactly the curious way that Western civilization, having ignored the canyon for centuries, abruptly veered into and through it. That cross-grained encounter was not the result of folk occupation but of determined exploration by intellectuals willing to refute common wisdom.

The canyon, Dutton asserted, was not sublime by virtue of any one feature, but "by virtue of its whole—its *ensemble.*" So it was with the era's exploration, which proved a vehicle not for any single perspective but as a means to assemble many, from geology, landscape art, nationalism, and the careers of critical personalities to institutions into a new whole. In such ways, amid settler societies, exploration not only exposed new lands for colonization but helped create the meaning such societies craved. The Second Age gave not only the newcomers but their lands a second chance.[5]

8

Lost Horizons

"Professor," replied the commander quickly, "I am not what you call a civilised man! I have done with society entirely, for reasons which I alone have the right to appreciate. I do not therefore obey its laws, and I desire you never to allude to them before me again!"[1]

—CAPTAIN NEMO, IN JULES VERNES'S *TWENTY THOUSAND LEAGUES UNDER THE SEA*

When the 20th century opened, the Second Age was pushing into the last reaches of an unexplored Earth. New Guinea was an intricate microcontinent, only slowly unveiled; its highlands had to await a gold rush in the 1930s. The exploration of Amazonia appealed to outsiders like Teddy Roosevelt on the River of Doubt and Percy Fawcett searching for the lost City of Z, though such expeditions were eddies, personal longings, away from the main currents that had carried the age. The core action lay where exploration might yet lead to power, wealth, fame, and rule.

The exploration of Tibet shows the close alliance of trade, empire, and geographic discovery at a place difficult to reach and with a people that did not want contact with the outside. The exploration of the Arctic, mostly complete, climaxes with a trek to the North Pole, the complex surveys of successive corps of discovery thinned to a single, simple quest. The exploration of the Antarctic shows the storm surge of discovery breaking against a landscape so alien—so lacking in variety, so empty of indigenes, so strange—that the old methods fractured.

Not only was the style of exploration running up to and beyond its limits, but as it pushed further it brushed against the frontiers of purpose and morality. For intellectuals searching horizons for the meaning of their age, the end of unexplored geographic space could align with an end of time; the absence of blank spaces, to a world without escape. The inexorable laws of progress might be inverted into laws of decay, seemingly bolstered by the logic of the second law of thermodynamics, such that inevitable improvement might be flipped into unavoidable atavism. Even as explorers sledged to the North and South Poles, Joseph Conrad probed a heart of darkness in which new discoveries led not to wonders but to horrors. Jules Verne, for many the voice of an irresistible technological progress that could carry humans to the center of the Earth, circumnavigate the planet beneath the seas, and propel passengers around the Moon, concludes with heroes, mad and isolated, that defy the expansion of empowered humanity and retreat to sanctuaries in the deep or at the end of the world, utopias of solipsism. Rather than being unbounded in space and time, the new Prometheans find themselves unexpectedly chained by their society's very success. The epitome of the Vernean explorer, Captain Nemo, has no desire to deliver his discoveries to a world he has renounced as hopelessly tainted.

So robust an era took a long time to fade; the Second Age was as tangled in its endings as in its beginnings. But what the Great War did abruptly for European imperialism, the loss of new lands and a revolution in intellectual syndromes did more slowly for exploration.

If the Institute of Egypt reminds us that exploration was often an aide de camp of empire, that it could exploit violence and be used to sanctify it, the exploration of Tibet shows where that tradition could lead. The race to discover the origin of the Nile or to reach the poles became a race to Lhasa.

The geopolitical setting was what Kipling termed the Great Game between Britain and Russia. Britain wanted India protected against rivals and was willing to pursue a forward policy of exploring parties and military forays to shield possible entry from central Asia. In such a context exploration was a species of espionage, and the scientist a spy. Naturalists could protest that they were only doing the good cause of geology and botany. The fiction fooled no one, but it was convenient.

The prize was the passes that threaded through the immense range of mountains and plateaus—the Karakoram, the Pamirs, the Himalayas, Tibet—that traced India's northern perimeter. Russia used exploring expeditions to convert the informal knowledge of fur traders and Cossacks into formal learning and to establish an administrative presence, even a claim. India, after all, had been invaded far more often by land than by sea. The British countered by sending parties masquerading as traders or scientists to identify those possible routes and secure them. Until the potential for conflict was defused by the Anglo-Russian Treaty of 1907, the Great Game dispatched a stream of exploring parties, some the work of eccentric travelers, some diplomatic and military missions. There were paths and passes through the mountains, but they were few and forbidden, save for the occasional merchant and pilgrim. Tibetans did not want to be discovered.

For empires, to protect from geopolitical rivals and to project trade, and for individual explorers and questers of varied nationalities to achieve the "dream of Lhasa"—these were incentives enough for Russians and Britons and the stray wanderer to penetrate the Roof of the World. The most widely traveled Russian explorer, Nikolai Przhevalsky, got within 150 miles before Tibetan troops turned him back. The Swedish explorer Sven Hedin crossed the plateau twice in 1901–8 but never reached Lhasa. The man who did was Francis Younghusband, who arrived at the head of a military expedition.

An Indian army officer, born in India, Younghusband had blazed a new route from China to central Asia, participated in secret missions for the Great Game, and won a gold medal from the Royal Geographical Society, all by the age of 26. Worried that either Russia, which had ambitions toward Tibet, or China, which claimed nominal sovereignty over it, might seize the country and threaten India, Lord Curzon determined to secure Tibet under the medium of a trade treaty. In 1903 he dispatched Younghusband to lead the Tibet Frontier Commission, accompanied by a serious military force under Brigadier General J. R. L. Macdonald, that would begin negotiations and demark the boundary between the two countries. The Tibetans refused to let the party enter. In December 1903 the British crossed anyway.

For months both sides tried to avoid bloodshed, but the British would be satisfied with nothing less than reaching their goal and the Tibetans with nothing less than denying them. Meeting with resistance the British decided they had no choice but to fight their way through, and the Tibetans, no option but to resist. Despite the formidable fortifications and hostile environment, British firepower quickly prevailed—artillery and Maxim guns crushed matchlocks. On July 31, 1904, Younghusband entered Lhasa. The mission left in September 23, with a treaty in hand and Younghusband aglow with a spiritual epiphany. The hardened explorer, the officer-cum-diplomat, the knighted embodiment of imperial Britain, was revealed as a disguised mystic. The dream of Lhasa had, for Francis Younghusband, become tangible.

For all its Edwardian dash the enterprise ended badly. The British had strode in with resolve and limped out with no little remorse. The treaty was later rewritten, memories soured, and Tibet was eventually absorbed by the People's Republic of China. In the end, the exploration of Tibet had come at a cost that neither side found acceptable. The winner of the Great Game turned out to be China.

The Arctic, long probed along its continental shorelines and islands, ended in a race to the North Pole that concluded not in triumph but in bitter controversy. For the American Robert Peary the pole had become an obsession. The pole drew him toward it with the gravitational pull of a black hole. Five times he had set out and failed. On February 28, 1909, at the age of 52, he turned his dog sleds north for what would have to be his final attempt across the great gyre of sea ice, broken by ridges and leads, that is the Arctic Ocean. On April 6 he claimed to have reached the pole. He made the final run alone: he, by himself, would exult in the realization of his dream. And so, too, he would know the nightmare that followed.

A former colleague, Frederick Cook, reached America before Peary returned and claimed that he had first reached the pole. Cook was already guilty of an exploring fraud when he bogusly claimed in 1906 to have climbed Mount McKinley, the highest peak in North America, but he was personable and Peary prickly. Now, while Peary was still struggling to return, Cook promptly set out on a world tour, where he was toasted and celebrated as the first man to achieve the pole. When Peary returned, expecting honors, he was met with indifference and criticism. Once again, it appeared, he had failed.

Though Cook was eventually exposed, he denied Peary what the hardened explorer thought his due. Cook was a charmer; Peary, a bleak obsessive, was widely disliked. He was better suited by temperament to drive a chariot around windy Troy than deal with public relations in the America of his time.

But that controversy spilled into others. Critics asserted that Peary had not in fact reached the pole on his final, implausibly long run, that he misread his instruments, that he falsified his notes, that the prospect of failure was too enormous for him to bear, that he was a fraud. That argument still exists, probably never to be resolved to the satisfaction of all parties. But over the years, as Western civilization shed its empires and adopted more inclusive values, other controversies have bubbled up. There was Peary's relations with the Inuit, with whom he lived and whose technologies he adapted (or in contemporary phrasing, appropriated) and whose lifestyle he accepted. And there was his long-standing servant, Matthew Henson, an African American, who had accompanied him on his travels and who was also denied a role in that final rush. Instead, it was all Peary—Peary as the emissary of white nationalist America. All this tainted his bid for a singular, personal triumph. As the years passed he seemed less national embodiment than national embarrassment.

Efforts to correct an evident injustice, if not a fraud, followed. Even *National Geographic*, which had supported him and serially honored him for his achievement, revised its official narrative. Books boosted the Inuit and Henson, who was subsequently acknowledged by the Explorers Club and granted a grave at Arlington National Cemetery. What had seemed a climax to a Second Age, what had testified to a restless American enterprise that had carried explorers across North America and now to the North Pole, became a milestone in the decolonialization that followed two world wars.

Antarctica offered a critique of another sort. Not since the Great Voyages had voyages of discovery produced tales of daring and survival to equal those of the heroic age. But a closer reading suggests some qualifiers to what many observers regarded then, and many still regard, as the high-water mark of exploration.

The expeditions were designed as scientific surveys, in keeping with the tenor of the late Second Age, but their grand narrative needed a goal. The south poles provided it—the magnetic pole, the pole of rotation. But the actual race to the pole between Roald Amundsen and Robert Scott offered minor drama. Amundsen reached the pole readily, his party a model of efficiency; the pole was all he intended. The expedition was hardy adventure, not science. More intent on doing some science as well as sledging, Scott stumbled to the pole in January 1912, and then staggered, fatally, back. The drama was his death march on the Ross Ice Shelf. What controversy surfaced concerned how he handled the polar party, but this was an internal critique, not one that reflected on the larger culture. On the contrary, his letters to the public, written as death approached, are treasured for their nobility of spirit. They voice sentiments that sound off-key after the wreckage of the Great War, but they rang true at the time.

Moreover, they found an able chronicler in Apsley Cherry-Garrard. The moral geography of *The Worst Journey in the World* centers on three tales from the expedition. One is the account of the Crozier party that made a mad, appalling, midwinter trek to the site of a penguin colony to collect eggs that, it was assumed, would contribute to the evolutionary history of the penguin—exploration in the service of science. The polar party was another—exploration as adventure and quest. The third was the northern party, a group of six men who found themselves marooned in a tiny snow cave on Inexpressible Island for the winter. They were reduced to sheer survival, living off penguins and the occasional seal. They met no goal, did no science; they simply managed to live out their exile in grotesque conditions. Beyond their collective saga there was little Cherry-Garrard could say that they accomplished. (When he visited the Natural History Museum to see the emperor penguin eggs, a "minor custodian flatly denied that any such eggs were in existence or in their possession." Later, a Professor Ewart wrote a report that concluded backhandedly that "the worst journey in the world in the interest of science was not made in vain.")[2]

Then there is the saga of Ernest Shackleton. He had been a member of Scott's 1902 *Discovery* expedition, then determined to lead his own, the *Nimrod*, which would attempt to reach the pole; he made it within 97 miles before having to turn back. In 1914, launching as Britain entered World War I, he led the *Endurance* expedition, which would attempt a cross-continental traverse, passing through the pole. The ship was caught and crushed in the ice. A second party, approaching from the other side of the continent, laid down some supply depots, but lost three men. After a long ordeal Shackleton managed to extract the *Endurance* crew safely in one of the epic feats of exploration leadership.

That leaves the story of Australian Douglas Mawson. He had been with Shackleton's *Nimrod* expedition and stayed another year with two companions to make the first ascent of Mount Erebus and to reach the south magnetic pole. He declined an invitation to join Scott's *Terra Nova* expedition in favor of leading his own Australasian Antarctic Expedition, which operated out of Cape Denison at Commonwealth Bay. Five sledging-party surveys were dispatched, with Mawson leading the eastern party. Five weeks out one member, Belgrave Innis, fell through a crevasse, taking with him the best dogs, most of the rations, miscellaneous essentials, and the tent. Mawson and Xavier Mertz turned back to base, but deteriorated physically and mentally as their odyssey persisted and they fed sled dogs to the remaining animals and to themselves. Mertz finally suffered seizures, then a coma, and finally succumbed. Mawson man-hauled the sledge the remaining hundred miles, only to find that the ship had left ahead of the winter ice earlier that day. He wintered over with six companions who had chosen to remain in the seemingly forlorn hope that he might return.

The essential trait of the ice is to reduce whatever is brought to it, and with Mawson exploration went to its final minimum. One man, an environment of sky and ice, a place where the cycle of the day and cycle of the year are the same, a panorama of white ice with no color, no sound or movement other than the wind, no life—there is not much for science, art, or literature, all of which rely on at least basic complexity of contrast, comparison, and conflict, only the murmur in Mawson's head and the solitary purpose, the most elemental possible for creatures, to live.

In the end, Mawson's return was an epic saga of survival, a legend of exploration, and a feeble story of science. By the time he returned, Shackleton had extricated the *Endurance* crew, Scott's polar party was buried in the ice, and Cherry-Garrard had supervised the erection of a memorial jarrah-wood cross with an inscription from Tennyson. Like so much of Western civilization exploration became a casualty of the Great War and its equally great aftershocks. The Second Age passed away with a muffled hurrah in the trenches.

By the time it faded, exploring had become not simply a job one could hire out to, but a career, even an avocation. Two men, both professional explorers, both adventurers as heart, illustrate how the era concluded. Roald Amundsen completed some of the classic quests of the age, Roy Chapman Andrews found himself equipped and sponsored and with nowhere, finally, to go.

The Norwegian Roald Amundsen did little science. His talent was to get to difficult polar places: the journey was sufficient. He learned his craft as a member of the Belgian Antarctic Expedition (1897–99). He then completed the Northwest Passage by small boat (1903–6). He was preparing for an attempt on the North Pole, but when he learned that it had been reached, he redirected his efforts south, and returned to Antarctica. On December 14, 1911, he achieved the South Pole. After World War I he navigated the Northeast Passage (1918–20). In 1925 he adapted to new technologies and flew from Svalbard to the pole, and a year later with a dirigible made the first crossing of the Arctic. He died in 1928 on a rescue mission for a dirigible lost over the pack ice. Along with Fridtjof Nansen, another Arctic explorer and later diplomat, he became a symbol of Norwegian nationalism after Norway achieved independence in 1905. His accomplishments could have made a career for several men.

Roy Chapman Andrews, too, dreamed of being an explorer since he was a child and prepared himself accordingly. "I was born to be an explorer," he wrote. "There was never any decision to make. I couldn't do anything else and be happy."[3] He wanted to work for the American Museum of Natural History and eventually did. He joined the Explorers Club in 1908, shortly after it was founded. Meanwhile, he trained in the skills he needed, did a shakedown cruise to the Arctic, collected specimens, and recorded his encounters on film. In 1920–22 he traveled through Mongolia by horse and car, then a wild region unsettled by the upheaval in Chinese politics. That led to the great accomplishment of his career, a multiyear scientific exploration under the rubric of the Asiatic Zoological Expeditions that took him to remote regions of China and Mongolia from 1923 to 1928 and had him return with the fossil bones and eggs of dinosaurs. For his achievements he became president of the Explorers Club from 1931 to 1934, and after that, director of the American Museum of Natural History, and decades later, one of the models for Indiana Jones. His own nationalism took the form of some light espionage—the Great Game continued on reduced rations.

Yet his treks looked like field science done in dangerous locales; and shorn of a quest, the adventuring could look like posturing. An explorer needed a place to explore and a compelling reason to go there. In 1937 Andrews directed his ambition to the Grand Canyon. Two isolated buttes, Wotan's Throne and Shiva Temple, promised, in his words, "a lost world." Surely, new species had evolved amidst that profound isolation; a journey to the two buttes could complement Darwin's voyage to the Galápagos Islands. In reality, it was an exercise in survey science hyped into old-style reconnaissance exploration.[4]

But it was the world of the Second Great Age of Discovery that was lost. What time and space had made a later time and another space could unmake. Humboldt could not have catapulted to fame had he climbed the Flint Hills of Kansas or botanized from a dugout along the Rhine. An exploring age that had fought through the cataracts of the Congo and clambered over the passes of the Tien Shan was reduced to scaling the Kaibab limestone, resupplied by airdrops in metal milk cartons. The episode spiraled out of control as the popular press transformed hope into hype and plumped what was, in reality, an elaborate backpacking trip into a search for dinosaurs and the Land Time Forgot. In the end the expedition discovered no novelties; the same species claimed both the plateau and its mesa miniature. Instead, the trek looked like a low-budget exercise in reenactment. The Second Age had entered a deepening twilight, with Shiva Temple and Wotan's Throne catching final patches of alpenglow before the night covered all.

Between the time Andrews had returned from central Asia and the time he wandered over the canyon's rim, James Hilton published *Lost Horizon*. An airplane crash (the modern equivalent to a shipwreck) leaves its passengers in a Tibetan valley surrounded by all-but-impassable mountains (which is to say, an island) where an ideal society flourishes in a lamasery known as

Shangri-la (a utopia). The story of the adventure is told by someone who met the protagonist, Hugh Conway, and heard the tale from him after he sloughed off amnesia and before he (Conway) disappeared again, apparently in an effort to find his way back to paradise. It's a familiar formula wrapped, as so many were, in the exotica made possible by aggressive exploration. Conway echoes Francis Younghusband in his spiritual epiphany amid the mountains.

In a sense Roy Chapman Andrews was trying to return to a time, as Hugh Conway was to a place, that enticed beyond reach. It belonged to another era. It was a poignant quest, but one doomed to wander in search of what could not again be found. Shangri-la existed in the mind and heart, accessible to the spirit, not to explorers tramping ceaselessly to unveil the Earth's fast-dwindling, still-lost worlds.

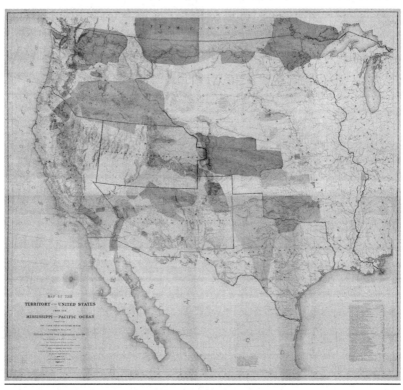

G. K. Warren map (1858). Lt. Warren of the Army Corps of Topographic Engineers included the results of the Grand Reconnaissance in antebellum America. Courtesy Library of Congress, Geography and Map Division.

The Apotheosis of Captain Cook. Details from design of P. J. de Loutherbourg, 1794; view of Karakakooa Bay is from a drawing by John Webber. Cook ascends to heaven assisted by Fame and Britannia, still clutching a sextant in his hand—a good example of the secularization of exploration, with Enlightenment science assuming the mantle of legitimacy previously enjoyed by religion.

Alexander von Humboldt in Ecuador, bringing Enlightenment to the Andes. Note the sunlight coming in from the right and the sextant Humboldt is showing the native. F. G. Weisch (1857).

Whole Earth photo from Apollo 17 (December 1972). Courtesy NASA.

BOOK III

Missions of Discovery

The Third Great Age of Discovery

There are no more unknown lands; the new frontiers for adventure are the ocean floors and limitless space.[1]

−J. TUZO WILSON, PRESIDENT, INTERNATIONAL UNION OF GEODESY AND GEOPHYSICS (IUGG) DURING THE INTERNATIONAL GEOPHYSICAL YEAR (IGY, 1961)

A historian of the twenty-first century, looking back past our own age to the beginnings of human civilization, will be conscious of four great turning points which mark the end of one era and dawn of a new and totally different mode of life. . . . [The fourth] and in some ways greatest change of all−the crossing of space and the exploration of other planets.[2]

−ARTHUR C. CLARKE, *THE CHALLENGE OF THE SPACESHIP* (1961)

Modernism Explores

The International Geophysical Year was to be different, not because it
would be bigger, though it had to be, but because it would be an attempt
to be all-embracing, to fit the earth into the pattern of the universe, to
relate its parts together, to discover hidden order, and to interpret the
whole in relation to space, and especially, to that great influence in nearby
space, the sun.

—J. TUZO WILSON, PRESIDENT, IUGG DURING IGY[1]

Once again, in a short period of time, three events triangulated the
contours for a new age of exploration. They were not so much separate expeditions as crystallizations of themes, all held together
within a common context, what became known as the International Geophysical Year (IGY) that ranged from July 1, 1957, to December 31, 1958.

In 1957 the International Geophysical Year assembled a swarm of investi-
gators from around the world to Antarctica, and while more symbolic than
scientifically productive, a Commonwealth Trans-Antarctic Expedition at
last traversed the continent. In October the Soviet Union launched the first
Earth-orbiting satellite, *Sputnik 1*, then a second; the United States answered
in January 1958 with *Explorer 1*. Meanwhile, British and American scientists completed tracing the Mid-Atlantic Ridge, at the time identified as the
Earth's longest ranging mountain chain, and a feature that even ten years
earlier had been unknown and unsuspected. While separate—each endeavor
aligned along a separate tradition—all operated under the loose aegis of IGY,
which had swelled from its origins as a polar year for Antarctica to embrace
the upper atmosphere, the ocean depths, and the Sun and Moon. Collectively, they announced a Third Great Age of Discovery.

Ice, space, abyss—two of those realms, ice and abyss, had frustrated and overextended the Second Age, and the third, space, had been denied it. Now, thanks to new technologies, scientific interest, and a keen geopolitical rivalry, they were poised to receive a reinvigorated wave of exploration, some by people, many by robots, all outfitted with instruments for remote-sensing, and all barren of the kind of encounters that had charged and defined discovery before. Intellectual culture, too, pushed beyond the pale of the Enlightenment into the peculiar realms and paradoxes of Modernism. Exploration was renewed: a heritage of geographic discovery that stretched back 500 years and had become moribund was revived. A Third Great Age of Discovery crystallized out of the slush. But as the Second Age differed from the First, so the Third Age would differ from the Second.

The International Geophysical Year was a catalyst, an announcement, a model, and a matrix for orchestrating ongoing and just-beginning projects. It effectively defined and demonstrated where and how a reinvigorated era of exploration could proceed. But it began with a return to Antarctica, where the Second Age had sledged to the end of what, for it, discovery had meant.

Antarctica was the transition, the icy pivot between eras. It was an abiotic landscape not much accessible to Enlightenment art and science, with no prospects for colonizing settlement, the last of the continental frontiers and the one where the Second Age had exhausted itself. Twice before—in 1882 and 1932—a global science community had rallied for "polar years." The earlier versions had focused heavily on the Arctic; this time, after a half century of world wars and dramatic technological innovations, proponents hoped to concentrate on the Antarctic.[2]

The scheme soon snowballed into a call for a more general 18-month scientific scan of the Earth. The roster of participants was a veritable United Nations of science—some 68 countries in all. IGY's explorers visited places inimical not only to humans but to life itself. They relied on remote-sensing instruments, rockets, submersibles, tracked vehicles, and robots. They inventoried planets whole, of which Earth was the prototype: the home planet became intellectually, a new world, the first of a dawning age of discovery that would propagate beyond the solar winds. The voyages that followed to planets such as Venus, Jupiter, and Neptune carried essentially the same instruments and asked the same questions of them as IGY did for Earth.

Through the infrastructure provided by IGY, noted J. Tuzo Wilson, the "science of the solid earth" was "absorbed into the broader framework of a new planetary science." Yet Earth's fluids interested the founders as much as its solids. They peered with special fascination into the upper atmosphere—geophysics, after all, was embedded into the project's very name. Auroras in particular had pointed to Antarctica as an insufficiently exploited platform for Earthly observation, which was where a third polar year had been headed. But the fast-morphing capabilities of rocketry made it possible to send instruments directly into the auroral belts. Both the United States and the Soviet Union had long-extant, if semidormant, plans to launch instrumented missiles beyond the realm of high-altitude balloons, leaving rockets as a new means to ask inherited questions. IGY gave them a goal and an incentive.[3]

It mattered that IGY bubbled up from scientists and that the International Council of Scientific Unions oversaw the agenda; individual nations contributed as they saw fit. IGY thus provided a neutral political arena, as Antarctica did a geographic one. For the major actors, the Cold War could play out on the ice under the guise of science, while for the smaller participants, it meant they could proceed without aligning with one side or the other. Territorial claims were ignored for the duration of the program. IGY allowed a geopolitical rivalry that expressed itself in a competition for prestige, not colonies; in public displays of prowess through launching satellites, not in seizing trade routes; and in ways that quarantined Antarctica and space from the unwanted contagion of a nuclear arms race.

IGY did for the Third Age what the voyages of Columbus and da Gama did for the First, and the geodesic expeditions to Lapland and Ecuador and the transits of Venus did for the Second. It resoundingly announced a phase change in exploration—a redirection in purpose, a chrysalis in style and perspective, and a retrofit of motivating enthusiasms. It held the pieces together in the force field of a new vision of a whole Earth. It established the Earth-Moon system as a proving ground: it was for this age what the Baltic Sea and the Greater Mediterranean were for the First Age, and a Europe remapped by science was for Second. Immediately after IGY, an Antarctic Treaty was negotiated that projected the operating conditions of IGY, notably open access, into the future. The Antarctic Treaty then served as a model for an Outer Space Treaty and a revised Law of the Sea.

Under the general aegis of IGY, the USSR launched *Sputnik 1* on October 4, 1957. It caused a global sensation and within the United States a political panic. The Americans had not thought they were in a technology race with the Soviets, had developed their own scientific rockets at a leisurely pace (*Vanguard TV3* blew up on launch), and could not immediately reply. The Soviets launched a second Sputnik a month later, November 3. The United States finally orbited *Explorer 1* on January 31, 1958. The terms of Cold War rivalry were on display for all to see. The contest was extended around the globe, and then to the Moon and beyond.

Like the three-stage rockets that carried most of the payloads, the exploration of space rested on three legs, the same that underwrote the Second Age. One was enthusiasm for colonization; a second, scientific research; and the third, the tradition of exploration as a cultural pursuit with roots that extended back half a millennium and had become integral to the identity of exploring nations. Those three motives did not align exactly: each had its own legacy and purposes. During IGY they synched perfectly. Later, as money and will weakened, they competed among themselves, and still do. But at the moment of creation, like separate compasses, they all pointed in the same direction. Within a dozen years the United States landed astronauts on the Moon. Within 20 years it launched the twin *Voyager* satellites that have passed through the solar system.

That left the deep oceans. Much as rockets replaced balloons in the upper atmosphere, so bathyscaphes, with a giant buoyancy chamber, supplanted bathyspheres, which had depended on cables to lower them. The *Trieste* was the prototype. Designed by Auguste Piccard, built in Trieste in 1953, operated by the French navy in the Mediterranean, it was then purchased by the U.S. Navy in 1958. On January 23, 1960, staffed by Lt. Don Walsh and Jacques Piccard (son of Auguste), it plunged into the Challenger Deep. Three hours and 15 minutes later *Trieste* rested on the bottom, roughly 36,000 feet below sea level. The *Trieste* was 7,000 feet farther below sea level than Mount Everest was above. The bathyscaphe was the exact analogue of the crafts being launched into space.

The descent came after IGY had formally concluded. But ocean exploration had geared up to serious dimensions in the postwar era, another phase of the Cold War. The war of course had stimulated interest in the geography of the oceans because so much naval warfare relied on submarines. The postwar era left the combatants with large navies ready for mothballing; a fraction was repurposed for exploration, in the United States under the Office of Naval Research. Ships trekked across the Atlantic, Indian, and Pacific Oceans trailing sonar, heat sensors, magnetometers, and other instruments. In 1957–58 Project Mohole was proposed to drill into the Mohorovičić discontinuity that separated Earth's crust from its mantle, a project that evolved instead into a comprehensive program of all-ocean core drilling. Gradually the voyages sketched the contours of the solid surface beneath the waters, which proved nothing like what scientists had thought it was. The ocean floors were a new world—the largest fraction of the solid Earth. Their study yielded a revolution in understanding how the Earth worked as a planet.

An old scenario revived, as once again the oceans helped birth a new age of discovery that even challenged what it meant to be an explorer.

As the Second Age entered its twilight, sledges were still the primary vehicle for traversing ice, balloons were the venue to reach high altitudes, and dredges and sounders dropped from ships the means to probe the deep oceans. All were further complicated by the need to accommodate humans: the sledges had to carry food, water, fuel, and tents; balloons, by the additional requirement to carry oxygen; and remote instruments, by the fact that they dangled from long cables under the control of human operators, though far from their sight and feel. Over the coming decades newly mechanized transport devices allowed further penetration and better control, though they continued to be burdened by the need to provide for human explorers who could not live off the land, hire native guides, or even breathe on their own. On the ice, vehicles had to be traveling habitats.

With government support, primarily military, airplanes, rockets, and submersibles were invented and then modified to serve traditional exploration goals. Previously, successful explorers had adapted equipment from natives; now, they drew from a war-catalyzed reservoir of inventions. Their complexity and the cost of carrying people, however, pointed to further mechanization, in this case of the explorer himself. Increasingly, robots complemented, and then competed with, humans as on-the-ground sensors and recorders. Of the many paradoxes of the new age, one of the most fundamental is that the price of getting people to uninhabitable places is that people are shed in favor of machines. This change of hardware, in turn, posed a challenge to the cultural software by which Western civilization understood the character of exploration.

The metamorphosis changed not just where and how exploration proceeded but its drama, its sense of human engagement and meaning beyond the gathering of data and images. Genuine exploration means more than extreme science or exotic adventuring. The exploration of places uninhabitable by humans without extraordinary built environments and prostheses redefined the character of the art, literature, metaphysics—the sustaining culture—that had made exploration an inexpungible feature of Western civilization and its sense of itself. The real discoveries would lie in how to make that transition in social software. The exploration of Mars would not resemble Humboldt or Lewis and Clark. The survey of the deep sea would not follow Jules Verne or an underwater Ferdinand Magellan. That was the route to parody.

But what should it be?

2

The Great Game Goes Global, and Beyond

The International Geophysical Year was conceived as the greatest attempt men have yet made to band together to examine, without passion or undue rivalry, their environment, their home and ultimate resource, the earth.

—J. TUZO WILSON, PRESIDENT, IUGG DURING IGY[1]

More pushes, more pulls. The old cauldron, Europe, that had stirred exploration and empire together and then boiled over across the globe now cooled. Rather than internal rivalry turned outward, the movement toward a European Union sought to calm competition among its members, even as an overseas imperium crumbled into decolonization. What replaced it was strife between the United States and the Soviet Union. Sputnik showed how the Cold War could give political performance enhancers to even nominally scientific experiments. The Cold War was the final propulsion module that boosted exploration across the ice, out of Earth orbit, and down to the ocean's floor.

In retrospect, the Great Game between the United States and the USSR lasted far less than those between Spain and Portugal, or Britain and France, but the era is young, and if it does in fact mark a Third Age, some other competitors, keen to secure national advantage or prestige through sponsored discovery, may emerge. While the International Geophysical Year provided an institutional matrix for science, there was plenty of passion and rivalry within its mission (Jock Wilson notwithstanding). Without the Cold War there would have been scant incentive and little urgency to erect bases on the Antarctic ice, scour the oceans for submerged mountain ranges and unexploitable trenches, or launch spacecraft and traverse the solar system.

Instead, two geopolitical rivals, both with active exploring traditions, chose to divert some of their competition away from direct confrontations and into proxy wars and surrogate contests, of which exploration into untrodden landscapes was particularly attractive. In this sense, IGY was the 20th century version of the transit of Venus project, in which France and Britain partly sublimated their global military confrontation under the guise of science. Power took the form of prestige. Returns assumed the form of information. Moon rocks replaced gemstones, and images of Saturn's moons, trade goods.

The Cold War was fought in the coldest regions accessible to the contestants, not only the white darkness of Antarctic ice, but the chilled emptiness of interplanetary space and the crushing blackness of the oceanic abyss. All relied on military vehicles, national security justifications, and a lightly weaponized science. Like the eras before it, the Third Great Age of Discovery was not merely a contest about knowing the Earth but a contest for it.[2]

The informing rivalry began, along with the Cold War, shortly after World War II. The United States and the USSR led the outpouring, but other countries like Britain, Norway, and Germany, with traditions of navy-catalyzed exploration joined, and especially France, which had Jacques Cousteau as a national champion.

The U.S. Navy, with hundreds of now-surplus vessels and with ample money ready to disburse through its Office of Naval Research, led the revival of American exploration. The creation of the National Science Foundation in 1950 added to the largesse for academic science. Of course the navy had military matters in mind—submarines were critical to naval warfare and were on their way to becoming a third of America's nuclear triad. Oceanography, as Jacob Hamblin notes, "was a Cold War science." Academic scientists did far better than 19th-century naturalists like Joseph Hooker and Thomas Huxley, hitching rides where possible on the post-Napoleonic Royal Navy.[3]

By 1948 the navy had invested $600,000 and donated two ships. Funding began to double routinely. A decade later the Office of Naval Research and the National Science Foundation contributed $25,000,000 a year—two orders of magnitude greater than what existed in 1941. By then Scripps had acquired an additional six vessels. From 1948 when it arrived, until it was sold in 1969, the R/V *Horizon* "sailed 610,522 miles all over the world on 267 oceanographic cruises for 4,207 days at sea." As Maurice Ewing of Lamont Doherty put it, "I keep my ships at sea." And at sea is where discovery occurred.[4]

The naval program extended to Antarctica. In 1946, nine months after Winston Churchill delivered his "iron curtain" speech, Task Force 68 sailed to Antarctica to secure a possible American claim, forestall Soviet ambitions, and test aerial photography and cold-weather maneuvering. Operation Highjump assembled an armada of 13 ships, including an aircraft carrier and submarine, along with 33 aircraft and 4,700 men; Robert Dietz wrote the oceanographic report. The task force left the ice a month before the United States announced the Truman Doctrine, and later the Marshall Plan. Highjump was followed by a smaller successor, Operation Windmill (1947–48), extending through the Berlin crisis. Then IGY provided an aegis for a continued presence.

And so it went, in a fugue of exploration and empire that stretched from the 20th century back to the 15th. The plans for IGY were formalized during the Hungarian uprising and Suez Crisis. In 1957, when IGY opened, Marie Tharp and Bruce Heezen published a physiographic map of the Atlantic Ocean's seabed, and *Sputnik 1* and *2* inaugurated the space race. In 1961 the Bay of Pigs invasion and Berlin Wall were matched by the first humans in space, Yuri Gagarin and Alan Shepard, and President John Kennedy's announcement that the United States would land a man on the Moon before the decade was out. The Cuban missile crisis paired with *Mariner 2*'s flyby of Venus. The submersible R/V *Alvin* launched in 1964; the U.S. buildup in Vietnam followed the next year. That same year both the USSR and the United States carried their Cold War competition to Mars. An Outer Space Treaty was negotiated through the UN in 1967; the next year the Soviets invaded Czechoslovakia. The Strategic Arms Limitation Treaty was signed as an updated Law of the Sea Treaty to govern the newly discovered seabeds was opened for signatures. During the period of detente between the end of the Vietnam war and the invasion of Afghanistan, the *Alvin* discovered hydrothermal vents along the East Pacific Rise and the *Voyager* spacecraft launched on their Grand Tour to the outer planets. When the Soviet Union dissolved in 1991, *Voyager 1* turned back from the edge of the planets and took its famous family portrait of the solar system, which included the Earth as a "pale blue dot."

Prowess in space especially was viewed as a test of competing political ideologies and economic systems, and the requisite hardware as demonstrations of military might. If you could launch manned capsules, you could guide missiles stuffed with multiple nukes; if you could prowl rugged terrain like the Mid-Atlantic Ridge with submersibles, you could maneuver nuclear-equipped submarines. Much as the Olympics were exploited for prestige, so was exploration. It also established viable claims to territory that might in the future turn valuable. Certainly the United States justified the size of its Antarctic presence on the grounds of national interests; and the Soviets had, during the hiatus offered by IGY, carefully placed stations in each of the Antarctic lands to which nations had announced territorial claims.

Without such commitments, it's difficult to see how exploration could have progressed so rapidly and so broadly. Expeditions that would have taken decades of normal science or plutocrat-financed forays now tumbled one after another. The crisis would come after the Cold War collapsed.

The Cold War provided a global push outward. The creation of the European Union pulled the traditional powers back. Since 1420 exploration and empire had been a two-cycle engine propelling European states around the world. The first half of the 20th century destroyed that motor. The second half labored to ensure that, for Europe, it would not be revived.

Even as the United States and USSR probed into terrae incognitae, Europe—at least its long-standing colonial powers—were leaving old ones. The big bang was the independence of India in 1947. The remainder of the British Empire, what had not been granted independence through more legalistic means (such as Canada, Australia, New Zealand, and South Africa) sloughed off in quick succession. France lost its overseas empire through oft-bitter fighting in Indochina and Algeria, and more peacefully elsewhere; but it was over by 1980. The Netherlands surrendered sovereignty to Indonesia in 1949. Germany's colonies went into UN trusteeship, then to independence. The scramble for Africa went into reverse, and with it the incentives that had propelled explorers into and across the continent. Instead, the Cold War replaced, for exploration, the former enthusiasms for trade, discovery, and conquest.

Even at the height of its imperium Europe had never been a unified presence. The endless squabbles of its tribes and later nations, and the repeated triumph of a balance of power doctrine that prevented any one country from dominating the whole, had kept continental Europe in endless conflicts and driven its internal rivalries to the ends of the Earth. After two world wars, largely fought on its own homeland, thoughtful statesmen argued that Europe could not tolerate another. The EU sought to bind its principle powers together in ways that would make another war unthinkable. That applied to its overseas dominions as well; Europe shed its colonies, some more readily than others, but dispatched them nonetheless. Paradoxically, as Alain Peyrefitte observes, "In point of fact, colonization was of little profit to the West, even in economic terms. . . . Those [countries] that lost empires with which they had lived symbiotically—the Netherlands, France, Belgium—experienced rapid economic growth precisely from the moment that they were relieved of their colonies. The richest countries of Europe—Switzerland and Sweden—never had any. The reason for this is that while trade brings profits, colonization eventually becomes costly." Colonization, he concludes, "was so foreign to the Western spirit that it was most often an unexpected result of unforeseen difficulties." The West wanted trade, and when that collapsed, colonization reluctantly followed, save for a brief fling prior to World War I. In recent decades Europe has achieved by a common market what historically it had sought by ferocious competition elsewhere.[5]

The emergence of the European Union is framed by some of the same key dates as the Cold War and Third-Age exploration. It emerges out of rubble of World War II. The Treaty of Rome created an Economic Community, expanding an earlier Coal and Steel Community, the same year as IGY.

The Maastricht Treaty (Treaty of European Union) followed the year after the Soviet Union collapsed. Membership steadily enlarged. Some of the old rivalries were often hard to suppress, however, and memories of previous empires difficult to slough off. In 1963 and 1967 France vetoed Britain's appeal for entry; in 2016 the UK voted to leave.

What matters is that the premise of the EU—"ever closer union"—worked against exactly those motivations that had sent Europe's quarrelsome peoples around the world. The EU sought to replace culture with commerce, military mastery with bureaucratic maneuvering, and brash exploration with disciplined science. The EU contributed to Third Age discovery, but through collective measures and as part of broad economic and scientific goals, not as an auxiliary to trade, conversion, conquest, or empire. Instead, the momentum for continued exploration transferred to the Cold War. When that conflict subsided, the great quest of the Third Age would be to find the incentive to continue.

The old collusions that had powered exploration yielded to new ones that calmed rather than quickened. If the European experiment sublimated conflicts that in the past might have erupted into war, it also sublimated contests that had once sparked rival expeditions, merging them into a collective science program. Wonder, not wealth, and prestige, not geopolitical power, moved Europe to contribute, mostly by way of scientific instruments, to the expeditions of the principal players. The informing rivalry had passed to its margins, east and west. When that quieted, observers worried that it might pass altogether.

3

Ice

One gets there, and that is about all there is for the telling.
It is the effort to get there that counts.[1]

—RICHARD BYRD, ON FLYING OVER THE SOUTH POLE

Antarctica—The Ice—remained both terminus and point of departure. Antarctica was, after all, a continent, reachable by surface ship,
sharing the Earth's atmosphere. Its perimeter could be treacherous
with moving ice and raucous blizzards, its interior abiotic but breathable; but
if it pushed the Second Age to its limits—"a continent of extremes and of
contrasts where there is no middle way," Jock Wilson observed—Antarctica
also offered entree for a Third. "It is in the Antarctic, which was least known,
that recent efforts have produced the most marked changes in our knowledge." The Ice held a historic pole of rotation between eras.[2]

Ships were mechanized—that's how parties negotiated the pack ice. The
heroic age experimented further with airplanes (Mawson) and tractors
(Shackleton), and even a balloon (Scott). But sledges pulled by dogs, Siberian
ponies, and men persisted as the means across the ice sheets and shelves and
to the pole. After the Great War, the airplane pioneered a new generation.
The scientific returns were greater, the hardships less; the character of Antarctic discovery slowly pivoted.

Oddly perhaps the momentum for discovery then passed to the United
States. America had not been part of the heroic age—its attention was riveted
on the Arctic, both the race to the pole and a gold rush in Alaska. U.S. Navy

pilot Richard Byrd changed that dynamic. In 1926 he flew over the North Pole and then determined to fly over the south as well. From late 1928 to 1930 he led a large party, including three planes and equipment for establishing wireless radio, to the Ross Ice Shelf and a base he called Little America. On November 29, 1929, Byrd and three others flew to the pole and back over the course of 16 hours.

From the heroic age to IGY, Richard Byrd remained the dominant figure in Antarctic exploration. He returned in 1934–36, another privately financed expedition, this time aligned with the second polar year. In 1939 the U.S. government wanted to establish a formal presence in Antarctica, and of course turned to now-Admiral Byrd. He oversaw two bases, east and west, in West Antarctica; they survived until American entry into World War II. When the U.S. Navy led America's return after the war, Byrd was again "officer-in-charge" of Operation Highjump (and its brief successor, Windmill). During the lead-up to IGY, Operation Deepfreeze, for which Byrd was titular head, he returned to the Ice one final time. In December 1955 he reached the American base at the South Pole established for IGY. His transition from privately financed discovery to the political science of IGY and the big-government operations of the Cold War tracks the course of Antarctic exploration. His extensive work in West Antarctica laid the basis for an American territorial claim, if the United States chose to make one, which it didn't.[3]

There were others. Sir Hubert Wilkens surveyed by air much of the coast of the Antarctic Peninsula (then claimed by Britain) during expeditions in 1928 and 1929. In 1935 Lincoln Ellsworth, an American millionaire fancying himself a polar Lindbergh, traversed the length of the Antarctic Peninsula from Dundee Island to Little America in a Northrup N-25, the climax to three attempts in six years. The British Graham Land Expedition launched the following year; then Nazi Germany flew over Dronning Maud Land. In 1946–47 Finn Ronne mounted a private survey along the peninsula. From 1949–52 Norway, Sweden, and Britain set up a joint base—the first truly international endeavor. In 1957 Vivian Fuchs, with help from Edmund Hillary, and with tracked vehicles resupplied by air, managed the first transcontinental crossing by land. By the end of IGY, exploration had segued into normal science, expeditions to permanent stations, and trekking by sledge to travel by snowmobile and aircraft.

The trend was clear: the old ways were overwhelmed by mechanical muscle, sophisticated instruments, state-sponsored surveys, and a globalizing politics that marked the transition from raw reconnaissance to systematic science. With remote sensing by seismic profiles and later microwave, field parties steadily penetrated through the ice and traced the lands beneath, much as similar techniques did for the ocean floor. Antarctica was brought more fully into the realm of earth science. It became an indelible if defiantly ineffable patch of Earth.

Antarctica offered more than geography, and needed more than science. It was a place. It had to connect to the culture if it was to share in the character of exploration. In this, too, it was transitional: it was the last, most tenuous expression of the Second Age, and the first, most challenging of the Third.

It came down to ice. Ice made Antarctica—literally. It blanketed East Antarctica and bonded that ice sheet to West Antarctica to make an ice continent. The freezing of the Southern Ocean each year doubled the size of the Antarctic ice field. The farther an expedition penetrated into the interior, the more ice it found, and only ice. Steadily, ice reduced and simplified what would otherwise have been a normal continent like Australia and a normal island arc of volcanoes like Indonesia into a single mineral, water, in its frozen, least mobile state, ice. The ice smothered everything else. There was nothing living save the explorer. There were no indigenous peoples. There was no society. Even seasons reduced to a single year-long day as the Sun spiraled around the pole. There was only the visitor and the Ice as Other. The quintessential Antarctic experience was a whiteout—ice underfoot, ice in the air, all distinction, direction, and perspective lost.

This is not the raw stuff that had, over half a millennium, made exploration so rich a source of information, adventure, encounter, and drama. If the whiteout was Antarctica's perfect geographic expression, the soliloquy was its literary equivalent, a white canvas its artistic representation, and the sheer will to survive its moral drama. The essence of Antarctica was not something added, not something that the Ice had that nowhere else did, but absence, things that should be present on a quotidian Earth and were missing or submerged in the transcendent Ice. That an ozone hole should appear above Antarctica is a perfect emblem of what makes it special. And what it does to Earth, it does to culture. Its power lies in its passivity. It takes, not gives. It removes, not adds. It reduces everything to a common minimum. It erases until, ultimately, it achieves an intellectual whiteout in solipsism.

There were plenty of places left on the Ice where no human had ever trod, on this, the fifth largest of the continents, but they looked little different from the places people had already visited. Future discoveries depended not on traverses by sledge, whether pulled by dogs, people, or snowmobiles, but by remote-sensing apparatus that did not even have to sit on the ice but could pass over sastrugi, crevasses, and whiteouts like a veil of cirrus cloud.

Part of what makes the heroic age so fascinating is the collision between the assumptions that had powered the Second Age and a landscape that offered no suitable reply. If a voyage to Antarctica echoed the classic passage to India, its icescape more resembled the Marabar caves in E. M. Forester's novel that only returned a single, common echo to every noise or query.

In Antarctica the core literature, the personal narrative, reduces the complexity and nuance of human life to a sheer struggle to survive. Fictional literature went further: it pushed beyond realism, however harrowing, into fantasy and science fiction, to a world beyond reason (think H. P. Lovecraft, *At the Mountains of Madness*) or to encounters with aliens (Don A. Stuart, *Who Goes There?*). So, also, art struggled to find ways to shoehorn what Antarctica presented into conventions of perspective, color, and meaning. The icescape was a setting better suited to the abstract expressionism of Josef Albers or Mark Rothko than the style of Romantics as Albert Bierstadt or J. M. W. Turner. The great photographs are of people—individual portraits of expedition members—rather than of places encountered. And of course there were no indigenes. The death of Belgrave Ninnis disappearing along with a sledge into a crevasse or of Captain Lawrence Oates removing himself from a tent into a blizzard to unburden Scott's polar party are dramatic, but have a different moral complexity from Ferdinand Magellan, sword in hand, dying in the Cebu surf or Captain James Cook killed by Hawaiians amid a cross-cultural confusion.

Antarctica had pushed the conventions of exploration and its cultural expressions beyond the pale, thinned them to the point of breaking. There were not many rocks to collect, no plants to gather, no terrestrial animals beyond the stray skua and the penguins on the shore, no natives to serve as guides or to challenge cultural assumptions. The high plateau that claimed inner Antarctica reduced terrain to a plain etched with surface sastrugi, geology to a solitary mineral, atmosphere to a washed-out sky or a black night dappled with auroras, as close to geographic nihilism as it was possible for humans to experience amid a breathable atmosphere. The scene was reduced to monosyllables: sun, moon, sky, ice, air, wind, cold. The usual tropes, measurements, protocols, and experiences were reduced to minima, stripped to an ice mirror that reflected back what the explorer brought.

Perhaps the ultimate expression was Richard Byrd's bizarre experiment during his second expedition in which he set up Advanced Base, a shack some 150 miles from Little America, where he proposed to spend the Antarctic winter by himself. It was as though he were creating artificially the kind of isolation and ordeal that had characterized the most compelling of Second-Age personal narratives but in reverse. Instead of striving, he would stay put. It was a curiously inert inversion of the sledging journey that that had been the staple of Antarctic exploration.

The reversal of plot ended with Byrd being rescued in midwinter. A faulty stove nearly killed him with carbon monoxide poisoning until a daring foray redeemed him. Henry Thoreau might seek a life of deliberation in a shack beside Walden Pond, which still allowed him trips to Concord for sustenance and society, but Byrd was as alone as it was possible for a human on the planet. The experiment had gone too far. To seek simplicity amid plenty can be a virtue; to seek it amid near nihilism can tip into vice. In his account of the experiment, Byrd concluded that "a man doesn't begin to attain wisdom until he recognizes that he is no longer indispensable." Richard Byrd may have been more transitional than he believed.[4]

So it also proved with geopolitics. Seven nations had asserted territorial claims, as though Antarctica could be parceled up as Africa had. But they were never able to follow up with permanent bases, much less colonies. On the Ice it was possible to keep the forms but not the substance of the Second Age. From the heroic age to the International Geophysical Year, Antarctica served as a kind of inflection point as exploration passed through a historical looking glass.

4

Space

The first voyage of men and women to Mars is the key step in transforming us into a multiplanet species. These events are as momentous as the colonization of the land by our amphibian ancestors and the descent from the trees by our primate ancestors.[1]

—CARL SAGAN, *PALE BLUE DOT* (1994)

Unlike ice or sea, space was not opaque, hiding new land beneath thousands of feet of ice or a handful of miles of saltwater. It was mostly empty and mostly just distant. Near-Earth space graded through an atmosphere that became more and more tenuous. A few mountains could carry intrepid adventurers to heights above 20,000 feet; elsewhere, and beyond that height, explorers had to rely on balloons and later on rockets.

The breakthroughs were the work of a Swiss physicist and inventor, Auguste Piccard, who developed a spherical gondola to carry people hauled upward by means of a helium-filled balloon. In May 1931, Piccard and Paul Kipfer reached an altitude of 51,775 feet, and in August 1932 Piccard along with Max Cosyns attained 53,153 feet. Before he ended his ballooning Piccard made 27 flights and reached 75,459 feet, over 2.5 times the height of Mount Everest. On each he recorded data relevant to cosmic radiation and other phenomena of the upper atmosphere. Then he demonstrated the adage that the mark of a genius was to have two great ideas when he redesigned the balloon into the bathyscaphe and pioneered the way to the deep oceans.

But even helium balloons were limited by the need for an atmosphere, however thinly extended. The future pointed to rockets, for which there was keen amateur interest and soon military necessity.

By World War II three national traditions of space travel flourished, each with its own cadre of partisans, each ready at the war's end to design and inspire a national program. One was European, one Russian, and one American. All proceeded at the same time more or less independently, all had origins in science fiction as much as engineering manuals, all tended toward the eremitic and the visionary, all were especially keen on the rocket for space travel, all became legendary in retrospect as the Cold War turned to space and the principal rivals searched their past for suitable patriarchs and champions.

The European was mostly German, but with French and British voices as well. (A French enthusiast, Robert Esnault-Pelterie, did pioneering work but failed to kindle a French tradition.) Its best-known proponent was Hermann Oberth (1894–1989), a peripatetic rocketeer, born in Romania, but whose professional career developed in Germany. He built a model rocket at the age of 14, defended his doctoral dissertation on rocket science in 1922, wrote his major works during the 1920s, culminating in *Ways to Spaceflight* in 1929, became a mentor to the Spaceflight Society (Verein für Raumschiffahrt), advised Fritz Lang on his wildly popular film *The Woman in the Moon*, developed liquid-fueled motors that powered Nazi Germany's rockets, and after the war worked in the United States under a former student, Wernher von Braun, and argued for the reality of UFOs.

The Russian tradition clustered around Konstantin Tsiolkovsky, a mostly self-taught recluse who helped found aerodynamics and addressed many of the fundamental problems of rocket propulsion in 1903 with his magnum opus, *The Exploration of Outer Space by Means of Rocket Devices*, before indulging a quixotic bent, arguing for a philosophy of panpsychism and for the necessity to colonize the Milky Way if humanity was to flourish. But he influenced Valentin Glushko and Sergei Korolev, and when the USSR committed heavily to rocketry after the war, he could serve as a Russian patriarch.

The American tradition orbited around Robert Goddard, inflamed by reading H. G. Wells's *The War of the Worlds*, followed by an epiphany in an apple tree in which he imagined travel to Mars. Another quasi-recluse, Goddard was less a visionary than others and conducted most of his experiments in liquid-fueled rockets through private philanthropy at remote Roswell, New Mexico. From 1926 to 1941 his small group, with philanthropic support, sent 34 rockets as high as 1.6 miles and devised much of the technology that would allow them to break the Earth's gravitational pull. Ironically, the U.S. military declined his services during World War II and built its postwar rocket program mostly out of German émigrés.

In their own time the rocketeers were largely dismissed as eccentrics, if not outright crackpots and utopians, inspired less by Isaac Newton than by Jules Verne, closer to Athanasius Kircher than Albert Einstein. They became celebrated in retrospect, elevated by the Cold War and honored by nations with a history of expansion and colonization, redeemed by their engineering skills when rockets became a practical and propaganda weapon of the Cold War and rivals sought to find national champions in their past. But their visions lofted their contribution beyond equations because the spacefaring nations had histories of expansionism and "space" assumed the burden of colonizing, and since there were no economic assets to be gained from trade or settlement, space travel became fraught with astrofuturism and outright millennialism that did not apply to the deep oceans or Antarctica. Exploration, especially in the service of science, was only one of several threads knotted into the space community.

The technological push came with World War II and the postwar Cold War. The German rocketeers were partitioned, like Germany, into American and Soviet spheres. Cult-like societies that propagandized for space as humanity's destiny suddenly found themselves in the vanguard. Scientists interested in the upper atmosphere turned to rockets, or to balloon-rocket hybrids, to carry instruments. Military enthusiasm for weaponizing missiles, particularly to transport nuclear warheads, boosted funding astronomically.

When, under the aegis of IGY, the USSR and the United States launched rockets, space became a public arena for Cold War competition. The Soviet Union made the first two successful launches, the United States the most launches. The United States discovered the Van Allen radiation belts; the

USSR photographed the far side of the Moon. The Soviets won nearly all the firsts; the United States had a wider and larger program. Before IGY ended, some seven nations launched rockets of some sort into the upper atmosphere. But what mattered was the contest between the two Cold War combatants. The unstated implication was clear: if you could put big payloads into orbit, or send them to the Moon or planets, you could also pack those missiles with warheads and land them where you wanted. But for many of the early rocket pioneers and their acolytes, the military was a means to build the technology that would enable people to migrate beyond Earth and colonize Mars.

The contest ran across two major environments. Near-Earth space soon filled with commercial and military interest—weather satellites, communication satellites, remote-sensing satellites, and reconnaissance satellites, along with orbiting astronauts and cosmonauts. In effect, rockets and their payloads acted as prostheses for extending traditional human projects beyond the highest mountains. Far-Earth space was more dramatic, and here is where the usual notions of exploration found their new worlds for discovery.

Most of the public and media attention focused on a "race to the Moon," a contest proclaimed by President John Kennedy in 1961. At the time the United States had lagged in virtually every space "first." The Soviet Union had launched the first orbiting satellite, the first animal, the first human, the first person around the globe, the first woman, the first multiperson crew, and the first machine to land on the Moon, and it published the first photo of the Moon's far side. The United States tried to reverse the tide by announcing that it would put "a man on the Moon, and bring him safely home" before the decade was out. It seemed one contest for a first-in-space that the United States might win.

The U.S. program played out before the media and the public. Every launch was a headline event; schools shut down classes to watch launches on TV (itself still a novel technology); like a serial drama, the events piled one on another, month after month, each leading inexorably to the Moon. The sequence of manned capsules grew with each iteration: Project Mercury, with one astronaut; Gemini, with two; Apollo, with three. Meanwhile, unmanned spacecraft, mostly the Lunar Orbiter series, studied and mapped the Moon. On July 20, 1969, Neil Armstrong and Buzz Aldrin landed Apollo 11's lunar module on the Sea of Tranquility. Five other landings followed, the last in December 1972, and one, Apollo 13 (1970), aborted and returned to Earth. The missions collectively brought back 837.87 pounds of lunar rock. The expression *moonshot* became shorthand for any monumental undertaking. But once the Moon was reached, and the mission replicated, the drama ended. When, during Apollo 14 (1971), Alan Shepard hit a golf ball in the Fra Mauro formation, it was as though Robert Scott, on reaching the South Pole, had decided to play cricket.

The manned space program shifted to near-Earth orbits. The Soviet Union sent its space station, *Soyuz*, into orbit with regular rockets, beginning in 1967. The United States responded with *Skylab* from 1973 to 1979. Meanwhile, it boosted the space shuttle as a general launch vehicle. The shuttle was intended to serve as a steady workhorse (it proved anything but), and crippled rather than quickened the overall American space program. Ultimately it endured two fatal explosions, one on liftoff and one on reentry, before being finally retired in 2011. Its stated purpose was to establish a permanent human presence in space through continually staffed stations. The costs of people in orbit, however, drove toward a shared International Space Station (ISS), assembled between 1998 and 2011. A weirdly circular logic, plus momentum, linked the shuttle and the station: the shuttle needed something to do, and the station needed a consistent means of supply. In the end Russian rockets proved a better bet.

All this kept people in near-Earth space, but not much human drama in the program, except where old space labs fell to a fiery end and shuttles blew up. Apart from the faltering shuttle and space station, however, there was plenty of activity: commercial and military satellites filled the skies, more all the time, from more nations, until so much debris floated around the planet that it has acquired a ring, and private companies emerged to furnish space tourism for the bored billionaires of global capitalism. It seemed a far cry from exploration. Even science fiction movie franchises like *Star Trek* found their plots veering from discovering unknown worlds into fighting familiar enemies.

The manned space race was an engineering marvel, a war refracted into expressions of technological prowess. The science came in the detailed studies of the lunar surface and then in returned Moon rocks. The exploration came through the narrative of putting someone on-site. But there was little that astronauts found that was not known, and little that machines could not have done in terms of geographic discovery. The lunar landscape had been extensively surveyed before Apollo 11 landed. The journey had been endlessly simulated. The whole point of such a costly undertaking was not to spark novelty, but to reduce the unknown as fully as possible.

The real action—the revival of geographic discovery—came through the exploration of the solar system. The closer the celestial body the more visits it had. The Moon had the most; then Venus and Mars; and the outer giant planets trailed off in proportion. Pluto was not encountered until the *New Horizons* spacecraft blew past it during the summer of 2015, and by then it had been demoted from the smallest planet to a largish member of the Kuiper Belt.

Planetary exploration proceeded in lockstep with the better publicized manned program. The Soviet Union's *Luna 1* flew past the Moon in January 1959; the United States responded in March with *Pioneer 4*. The Soviets sent *Luna 2* to the surface in September, and *Luna* around the Moon in October. In December 1962 the American *Mariner 2* successfully conducted a flyby of Venus. The robotic race to the planets was on.

What followed has been called the golden age of planetary exploration. Between 1960 and 1989 the two rivals launched 36 missions to Venus and 30 to Mars. For the USSR, spacecraft to Venus succeeded in 14 out of 29, and for the United States, all 7 met goals. For the USSR's spacecraft to Mars, only 2 out of 20 succeeded, and for the United States, 8 of 10. Most Soviet failures occurred in the early years, and most involved Mars. Beyond the inner, Earth-like planets America sent *Pioneer 10* to Jupiter and *Pioneer 11* to Jupiter and Saturn. In 1977 it launched the twin Voyagers—*Voyager 1* to Jupiter and Saturn, and *Voyager 2* to Jupiter, Saturn, Uranus, and Neptune. Both are still functioning; *Voyager 1* entered interstellar space in August 2012, and *Voyager 2* in December 2018.

Overall, U.S. spacecraft proved more robust than their USSR rivals. Unlike the Soviets, Americans did not design for future human crews; they could miniaturize, they could simplify. American success in electronics paid off in guidance systems and more compact payloads. Besides, mission failures—and there were many on both sides—did not result in human deaths. They might dishearten and demoralize political enthusiasms, but they were not likely to decommission programs. Compared with the complexity and cost of human flights, two spacecraft could be dispatched on a common mission, giving a useful redundancy to the project.

The Voyagers completed the space equivalent of the carreira da Índia. Subsequent missions around the solar system targeted the Sun and specific planets and their moons for long-term surveys, while Mars missions segued into field science, answering questions about climate, water, and the ever-enticing question about the prospects for life. Taken together, the range and speed of planetary discovery—its scope, its vigor, its payoffs—compare well to those great outbursts of the Iberians in the First Age and of settler societies in the Second.

All those who participated in the American planetary program from *Mariner 2*'s flyby of Venus in 1962 to Voyager's embarkation in 1977 agreed that this was a privileged time. Spacecraft visited every planet, and revisited the closest; almost every year saw a mission, and when, after the hiatus imposed by the space shuttle *Challenger* tragedy, planetary exploration revived, it built on the legacy and hibernated ideas of those epic years. It was all "the stuff of legend and myth," as space historian Roger Launius put it. Proponents differed only in their sense of the era's tempo and their reckoning of its capacity to persist. They fretted because the age depended less on engineering cleverness and scientific purpose than on the whims, wealth, and mores of its sustaining society. Voyager, Bruce Murray noted, was the last mission in which "the technical challenge was dominant. Since then it's been politics." The golden age segued into a silver one.[2]

Of the three arenas of the Third Age, space—for good and ill—was the most encumbered with cultural legacies and futuristic fantasies. It had the lure of the night sky—even naked eyes could see other suns and planets and what were later understood to be galaxies and subuniverses and evoked marvel about humanity's place in the great scheme of things. Other worlds invited visions of colonization, nurtured a literature, lured with a Siren's song.

It was colonization, the prospect to encounter new Earthlike worlds, that distinguished space from ice and abyss. It was easier to imagine spaceships voyaging through a vacuum to a new-world Mars than vessels plunging to the crushing depths of the ocean to submarine civilizations. A century of contact with Antarctica had not led to colonies, but at best to bases staffed by a rotating population of scientists with logistics mostly supplied by military vessels—a Third-Age version of First-Age trading factories. Efforts to colonize the oceans started boldly, then limped away from the future. No TV or movie series focused on Antarctica, and none on the ocean survived. They were not only too harsh but too close. The unfathomable distance of space allowed the imagination to project what it wanted. But even Carl Sagan, who insisted that "the zest to explore and exploit" had "clear survival value" and was common to all of humanity, admitted that "Earth is where we make our stand."[3]

"Space" was a complicated idea and a syncretic community. When America launched its first successful satellite, three contributors to the project raised a mockup of *Explorer 1* at a press conference. Together, they stood, literally, for the three technologies and the three purposes that had converged to support the endeavor. Wernher von Braun built rockets and sought off-Earth colonies. James Van Allen built instruments and sought off-Earth platforms—labs—for doing science. William Pickering, director of the Jet Propulsion Laboratory (JPL), represented the institution that assembled the package and oversaw its operations, and he sought to define the program as part of a grand tradition of exploration. In 1958 they all contributed—as did, sub rosa, the military since space was politically reckoned as part of national security.

But those three goals were not identical, nor always compatible. Colonization demanded people, and colonizers considered technologies only valuable so far as they supported the peopling of new worlds, of which Mars seemed always the obsession. Scientific inquiry did not need people in space: it needed instruments that gathered and beamed back data. It could do that from stationary platforms like the Hubble Space Telescope. Exploration did not have to lead to colonization, and it did not need science as a goal. The

First Age had exploded across the Earth without modern science, and it had largely shunned colonization, which was expensive and complicated, in favor of trading posts. What exploration demanded was a trek and a narrative that involved a quest beyond simple adventuring. The convergence of all three ambitions had occurred during the Second Age and seemed to casual observers part of an indissoluble whole. In its space realm the Third Age began with all those notions bundled, but it could not sustain them equally. As costs and challenges escalated, so did the internal tensions among the three goals. They began to split.

The fulcrum was the human presence. People were expensive to send into space. They complicated missions, and they savaged budgets. But for those advocates for whom space only had meaning as an arena for colonization, the program was about sending people to other planets, in the end permanently. People didn't have to be in space to do science, however, and strains developed between the colonizers and the experimentalists. The issue went public when, on the last day of *Voyager 2*'s encounter with Uranus, the space shuttle *Challenger* blew up on launch. The wealth of data and dazzle of discovery stood in stark contrast with the risks of human spaceflight—and their meager returns on science.

But neither did science need journeys except to position its instruments. The Voyagers discovered 26 new moons among the outer planets; by the time *Voyager 1* left the heliosphere the Hubble Space Telescope had found 48 around Jupiter alone, most of them the size of asteroids. If data was the goal, experiments located in space could replace expeditions. A trek was just something done to get the lab in position. The journey was one of the mind, not of corporeal bodies.

The separate factions could adhere with the strong nuclear force of money and politics. Every partisan tribe could get what it wanted. When the funding faded and the politics turned to other concerns, trade-offs had to be made. If humans in space stirred the noisiest interests, it also provoked the greatest costs. The space shuttle had stymied, and nearly killed, planetary exploration, and when it finally degenerated into just old technology, its legacy was to leave the United States without a suitable launch vehicle. The ISS continues to struggle for a commanding purpose. Mars visionaries look to the private sector.

The robots were klutzy compared to humans, but they were adequate for reconnaissance discovery, and they were evolving at exponential rates. People were not. The colonizers just became louder. What space lacked in terms of expeditions it made up for in literature and film. Space could accommodate imagined colonies and encounters better than ice and abyss. Science fiction and technological romance flourished in print and on screen, and in a succession of cheerleaders from Arthur C. Clarke to Carl Sagan, the Edens and Hakluyts of the Third Age. Even robots could become sentient, allowing for moral drama to fill the void. Only by going to Mars could we save Earth. Only by going into deep space, even if it meant genetically engineering humans into beings capable of long-duration space travel, could we save humanity. Echoing St. Paul, it seemed that the price of immortality was to overcome the natural man.

By the time the Voyagers actually crossed into interstellar space, the champions were gone, and the urgency dissipated. Under human meddling Earth itself was changing faster than our understanding, much less our ability to manage. We needed less to look out in wonder than to look back in worry. The enduring image of the early space era was not the Moon or Mars, but the Earth whole, the Earth rising over the rim of the Moon, the Earth as blue gem, precious and unique, set in the velvety blackness of empty space.

5

Abyss

We felt as if we had glimpsed unknown, alien life on a new world, or at least an alternate version of life on our own.[1]

—ROBERT BALLARD, ON THE DISCOVERY OF BLACK SMOKERS ON THE EAST PACIFIC RISE

Europe was a fractal patchwork of peninsulas and islands surrounded by seas. Its waters fed its peoples, carried its trade, dispersed migrants and warriors, and sustained its empires. Its threats came from the Eurasian plains to the east; the oceans offered a sanctuary, an escape valve, and an opportunity. When Europe could not satisfy its ambitions landward, it turned to the sea.

The oceans had led each Great Age of Discovery. The world sea had been the special arena of the First Age; the mapping of its littoral, its special achievement. In the Second Age the ocean had carried explorers to the continents and new islands and had been subject to long-running surveys that brought the Enlightenment to a watery Earth. Matthew Maury had applied geographic principles to the sea as Humboldt had to land. The *Challenger* expedition had combined science and geographic discovery in much the same way the Great Surveys of the American West had, and like them had competed with rival undertakings from Norway, Germany, France, Russia, and Austria. By 1902 an International Commission for the Exploration of the Sea was established, followed by a general bathymetric chart of the oceans. In 1910 Monaco opened an Oceanographic Institute and museum.

But the ocean was difficult to sample much below the surface. Cold, black, void of oxygen, dense with crushing pressures—it was impossible for conquistadores or marinheiros to ride or sail below the waves, and shipborne dredgings, nets, and soundings on cables were slow, cumbersome, and laughably few compared to the immensity of the deep. It was possible to climb mountains; it was not possible to clamber into ocean trenches. There were few rivals since there was little to compete over, scant means by which to do it, and few incentives. The abyss was Antarctica without oxygen, light, and a surface to sledge across; it was space without stars to prick the blackness, with pressures as dense as vacuums were empty, with a cold close to absolute zero. "The history of the exploration of the sea floors is brief and simple," Wilson intoned, "and it can truthfully be said that the enormous task of studying these cradles of the sea was first faced in an adequate manner during the IGY.[2]

But the deep sea had features that ice and space lacked. In its sediments it held revelatory information about Earth history, in its rifts and ridges it told the story of planetary dynamics, and even in its most oppressive depths it held life. The abyss rather than space would dominate the Third Great Age of Discovery, and notwithstanding Wilson's understanding, even before IGY it led the great wave of postwar exploration.

War and individual pluck changed the conditions. Submarines altered the strategic basis for naval warfare. Navies now needed maps of the sea floor, means of subsurface communication, and basic oceanographic geography; they needed to know Earth's oceans as air forces did its atmosphere. Mostly, progress was technological—better submarines and the means to combat them. All this accelerated through World War II and into the Cold War. After both wars surplus naval surface ships became available for oceanographic science. Until then two inventor-adventurers pioneered vessels to carry observers into the ocean depths.

The American William Beebe began in the tradition of the explorer-naturalist. In 1899 he became curator of ornithology for the New York Zoological Gardens, and in 1919 director of tropical research for the New

York Zoological Society. He visited Mexico, Trinidad, Venezuela; in 1909 he traveled the world to research pheasants; later, he toured Brazil, established a research station in British Guiana, explored the Galápagos Islands, and worked around the Caribbean and Bermuda, where he set up another research station at Nonsuch Island. He wrote up his travels in wildly popular books.

His Bermuda days inspired him to direct his collecting treks and natural-history cross-sections to the offshore seas. Eventually he attracted the attention of Otis Barton, an engineer, and together they devised the bathysphere, which could be lowered by cable and came with electricity and telephone. From 1930 to 1934 Beebe and Barton made dives, to great public enthusiasm and publicity. The dives had all the apparatus of traditional exploration, and were received as such. Having found what he wanted, Beebe then returned to tropical zoology.

By then Auguste Piccard realized that the high-altitude balloons he had devised could be adapted into vehicles to plunge down into the ocean. In 1937 he designed the bathyscaphe; in 1945 he began construction on the *FNRS-2*. Redesigned, it became the *Trieste*, launched in 1953 by the French navy for work in the Mediterranean Sea. Those early dives inspired Jacques Cousteau, then a French naval officer working with the *Trieste*, on a career that made him the public face of oceanography, and they led to the purchase of the *Trieste* by the U.S. Navy.

The United States acquired the *Trieste* as the International Geophysical Year was in its final phase. Satellites effectively announced space exploration, moving beyond the upper atmosphere; submersibles similarly advertised entry into the deep ocean. For both IGY provided a context. For space it served as a kind of public launchpad. For the deep oceans, it acted as a way of incorporating into a planetary matrix the remarkable discoveries made in the immediate postwar era.

Not much was known. By 1914 only 1,800 soundings had been made in north Atlantic, "and fewer elsewhere." Only three ships had been built specifically for oceanographic research, all destined for the poles (Nansen's *Fram*,

Peary's *Roosevelt*, and Scott's *Discovery*). A rough bathymetric chart of the oceans had been published. Mostly the seafloor was assumed to consist of vast plains of ooze. Some terrain had been identified—plateaus, the occasional seamount, and the Challenger Deep near Guam. Even in the 1950s Henry Menard recalled the prevailing vision as one of an "old and relatively static world. No beginning, no end, slow erosion and sedimentation, changes gradual, and so on." It was worse in the oceans, which were seen as being "as old and even more static than the continents." It was, he concluded, "pretty unlikely prospects on which to build a professional career." The Moon was better mapped.[3]

Then the postwar era liberated a storehouse of instruments, funds, and vessels, and sparked an energetic purpose. In America the Office of Naval Research repurposed surplus ships and poured money into oceanography. Other funding appeared from the newly endowed National Science Foundation. Two research centers emerged, one for the Atlantic Ocean at Lamont-Doherty Observatory at Columbia University and one for the Pacific at Scripps Institute of Oceanography. Their fleets ran cross-sections over the world's seas that cored sediments, measured heat, geomagnetism, gravity, density, and mapped the floor with echo sounding. By 1956 those efforts, combined with improving maps of earthquakes and submarine volcanoes, pointed to an immense rift and range halfway across the North Atlantic. A year later, as IGY opened, Tharp and Heezen published their physiographic map of the north Atlantic.

With IGY as an institutional matrix, "the tracing of this vast system" of continental and mid-ocean "fractures," Wilson noted, "was diligently pursued, and the discovery that it is continuous was one of the most notable achievements of the IGY." Or as Menard recalled, "after the global viewpoint provided by the IGY, everyone realized that whatever was done on a grand scale in one region had to be balanced by some reverse phenomenon on a similar scale elsewhere." In brief, the oceans were absorbed into the Cold War, and IGY redirected that rivalry in ways that allowed other nations to contribute and that jolted the field with money. During IGY two American submarines, the USS *Nautilus* and USS *Skate*, surfaced through the Arctic pack ice at the North Pole, announcing an American presence to match the establishment of an American base at the South Pole and showing the capabilities of submarines to operate in the Arctic as Sputnik and Explorer had demonstrated the capacity of rockets to reach space and orbit the Earth.[4]

This was publicity and posturing. What helped perpetuate the IGY experience was the creation of a Special Committee on Oceanic Research, much as a similar program did for Antarctica and a comparable one for space. In 1958 the idea arose for drilling into the Mohorovičić discontinuity that divided the Earth's crust from its mantle and returning a sample; the drilling would occur in the oceans, where the crust was thinnest and where off-shore drilling techniques were rapidly improving. Project Mohole was a moonshot—and to critics another "Moondoggle" in the making. It got a hearing and preliminary funding from 1958 to 1966, but not final approval. Meanwhile, under the auspices of the Special Committee on Oceanic Research, an International Indian Ocean Expedition mobilized 40 ships and scientists from 20 countries. Those mission-specific treks were followed by an International Decade of Ocean Exploration (1971–80). What, in Wilson's estimate, had been the "least known" ocean would soon become among "the best known."[5]

There were others who helped make it so, probably none so much as Jacques Cousteau, a French naval officer and co-inventor of the Aqua-Lung. In 1950 he converted a Royal Navy minesweeper into the research vessel *Calypso*. With support from the French navy's Undersea Research Group, funding from the Centre National de la Recherche Scientifique, which sponsored oceanographers for the ship, and work from commercial sources like British Petroleum, Cousteau and *Calypso* became the face of ocean discovery, an endless voyage around the globe, emphasizing the near-surface ocean, down to a thousand feet or so, the equivalent of near-Earth space. In 1962 he installed the first of a series of Continental Shelf Stations (Conshelf), a quasi-permanent habitat from which divers could explore the sea around them. A gifted photographer with an instinctive sense for popular taste, Cousteau recorded the glamorous realms of coral reefs and the boisterous life of fish schools, sharks, and squid. The role Carl Sagan later assumed for space exploration, Jacques Cousteau pioneered for the oceans. *The Undersea World of Jacques Cousteau* was a rough model for Sagan's *Cosmos*, not least the central role granted the presenter.

No less significant, nature writing found a new world in the ocean. In 1947 Ferdinand Lane published *Mysterious Sea* and R. E. Coker, the more scientific *This Great and Wide Sea*; four years later Rachel Carson followed with her best-selling *The Sea Around Us*, whose adaptation by RKO into a film documentary won an Academy Award; four years later Cousteau's founding documentary *The Silent World (Le monde du silence)* took the Palme d'Or at the Cannes Film Festival and then an Academy Award. Tuzo Wilson's pronouncements were largely true for science, but public expectations for ocean discoveries were well established before IGY launched and Sputnik redirected attention to space. Arthur C. Clarke's *The Exploration of Space* followed Carson's *The Sea Around Us* by a year.

Over the next decade collaborative oceanographic expeditions splashed across the world oceans. The International Decade of Ocean Exploration segued into the Joint Oceanographic Institutions Deep Earth Sampling program (JOIDES) with the drilling ship *Resolution*, while Mohole morphed into the Deep Sea Drilling Program that from 1968 to 1983 took samples from around the world ocean and, along with geomagnetic and topographic readings, helped revolutionize prevailing understanding of Earth history and dynamics. Individual nations with an interest in oceans, galvanized by the collective discoveries, added their own contributions, further leveraging the whole.

All together the expeditions mapped the anatomy of the world ocean and measured its physiology. They discovered the world's largest mountains, its tallest peaks, and its longest ranges. They found Earth's most profound gorges. They worked out the deep as well as the surface circulation of its waters. They collected thousands of sediment cores that testified to Earth history otherwise lost on continents. They amassed data on heat flow, gravity, seismicity, and magnetism. Within the world ocean they traced the grandest migrations (even of whole ecosystems) and found exotic niches for life. They revealed the oceans as not passive basins filled with terrestrial wash-off and zoological detritus but active engines that moved continents and hosted biotas based on novel biochemistry. They found new answers to the erstwhile question of why the sea is salty. One of the goals of the International Geophysical Year had been to reimagine the Earth whole. No part contributed more to that vision, and to the conceptual reformation that followed, than did the exploration of the ocean depths.

All this fervor, however, remained embedded in the traditional vehicle for ocean travel, the surface ship, now outfitted with remote-sensing capabilities. The truer transition, the equivalent of spacecraft, was the development of submersibles.

For the United States the acquisition of the *Trieste* by the navy was a point of inflection. The navy sought not only information about its primary environment but vehicles suitable for rescue operations and capable of searching for lost ships and in one notorious incident, hydrogen bombs accidentally dropped from a B-52 off the coast of Spain. The reconfigured *Trieste* 2 launched in 1964, the same year the navy commissioned what became the most celebrated vessel of the fleet, the *Alvin*, under the direction of the Woods Hole Oceanographic Institute. Over the years the *Alvin* underwent incessant reincarnations and repeatedly demonstrated its unexcelled value. By January 2019 it had made 5,000 dives.

It was initially a hard sell for scientists. Like astronomers who wanted space telescopes more than space treks, oceanographers understood what surface vessels could do, and didn't want upstarts like submersibles to compete and perhaps siphon away funds and interest. For marine science *Alvin* proved itself during the 1973–74 Project FAMOUS expedition, a joint United States-French campaign that toured the Mid-Atlantic Rift Valley. As Robert Ballard recalled, "I remember all too well the substantial finger of Maurice Ewing," a strong champion of traditional oceanography, "wagging in my face . . . threatening to have *Alvin*'s pressure sphere melted down into paper clips if it failed in this mission."[6]

It succeeded, and went on to a succession of triumphs. Scientifically, its most celebrated discovery occurred during the Galápagos Hydrothermal Expedition, along the East Pacific Rise. In February 1977 the *Alvin* and its camera sled, *Angus*, stumbled into a patch of hot vents, or black smokers, that supported sulfide-oxidizing bacteria, giant clams, something that resembled yellow dandelions, "along with white crabs, limpets, small pink fish, and clusters of vivid red-tipped worms that protruded from stalklike white shells, or tubes." It seemed to Robert Ballard as though he stumbled upon an unknown alien world, a road otherwise not taken in Earth's evolution.

The discovery became one of the epic moments of the Third Age. Politically, submersibles remained fundamental to the mission of the navy. Publicly, however, *Alvin* achieved celebrity when in April 1986 it found the sunken *Titanic*—followed by other notable shipwrecks.[7]

The civilian programs were known, often promoted in venues like *National Geographic*. *Alvin* was only the brightest celebrity, and by being continually upgraded into newer avatars, it retained its visibility. The most dramatic developments, however, were black programs whose full accomplishments did not become public until the end of the Cold War and the opening of relevant archives. After 1991 a slew of submersibles appeared, from multiple countries, for varied purposes. The post–Cold War submersibles served a similar role as surface vessels had after World War II. Scientific revelations tumbled one after another. Then autonomous submersibles were devised that could explore on their own, and then packs of vessels, veritable swarms, that could communicate with one another—new suboceanic corps of discovery. Increasingly, robots replaced human explorers. Ballard, who had pioneered both, thought the robots were better. After telepresence became possible, he exulted, "Our minds can now go it alone, leaving the body behind."[8]

But after the great postwar surge, even with a second boost from the conclusion to the Cold War, raw reconnaissance and discovery followed the classic trajectory toward routine field science. Other priorities competed for national attention, rivalries among oceanographic institutes replaced that between Cold War antagonists; the fever cooled. Much as space programs pulled back to near-Earth orbits, looking for commercial, military, and scientific projects, so deep ocean exploration turned to environmental concerns, military needs, and near-continent shelves. Still, oceanic exploration flourished—the most vibrant and productive arena for Third-Age discovery. "From 280 meters downward," as Jacques Cousteau exulted, "everything is to be discovered."[9]

Space and abyss continued in an eerie fugue. Bathyspheres were the high-altitude balloons of the sea; the bathyscaphe *Trieste* descending to the Challenger Deep was the sibling to *Mariner 2* flying past Venus; the DMS *Alvin,*

to the STS space shuttle; and there was a comparable shift from crewed to robotic probes. Project Mohole was an Apollo program for the deep ocean, and the Deep Sea Drilling Project made the transition to routine coring in ways the space shuttle could not routinize space flight. SEALAB was the abyssal avatar of *Skylab*. Ambitions to colonize the ocean floor were sprinkled with the same pixie dust as Gerard O'Neill's orbiting colony at the Lagrange point between Earth and Moon. Aquanauts competed with astronauts; NOAA was boosted as a wet NASA. Schemes to gather minerals from the ocean floor ran shoulder to shoulder with proposals to harvest asteroids. There was even precedent in science fiction; after all, Jules Verne had written *20,000 Leagues Under the Sea* as well as *From the Earth to the Moon*, and Arthur C. Clarke published novels about the sea that complemented those about space: to his *The Challenge of the Spaceship* he added *The Challenge of the Sea* (1960), which came with an introduction by Wernher von Braun. The Cold War rivalry played out beneath the waves as fully as it did beyond the clouds. By the end of the 20th century both realms felt the almost geologic power of humanity as Earth acquired a Saturnian ring from orbiting space debris and the oceans sported mammoth gyres of plastic. Yet while the Earth's oceans were more accessible and had more practical meaning for humanity than anything in space, space held a mystique that the abyss lacked.

The deep oceans drifted from public attention. Secrecy is one reason; much of the oceanic work had military sponsorship and was not publicized. Nor was there any overt, broadcast-over-TV, soap-opera drama for an abyss race as there was for a space race. There was no NASA to oversee launches, no countdowns on television. There was no literary legacy for deep-sea exploration as there was for planetary. Robert Heinlein did not write a novel titled *Submarine Troopers*, nor Ray Bradbury a *West Mariana Basin Chronicles*. Arthur C. Clarke did not imagine *Childhood's End* happening on the East Pacific Rise. Stanley Kubrick did not film *2001: A Sea Odyssey* or place alien obelisks on the Valdivia Abyssal Plain. There was no Tsiolkovsky to project a Great Migration to the Laurentian Abyss, or a Percival Lowell to sketch the contours of a dying civilization on the Loihi Seamont. There was no Carl Sagan to fantasize about cosmic connections with galactic intelligences, or rhapsodize about chemo-spiritual liaisons with "salt stuff" shared between people and black smokers; over the years Jacques Cousteau seemed more showman than scientist (the newsmagazine *L'Express* quoted one French savant that Cousteau was to oceanography what the cancan was to *Swan Lake*). Arthur C. Clarke's sea novels sank; his space fiction soared. The deep attracted little imaginative literature; Michael Crichton's *Sphere* succeeded in part because it combined space and abyss. (Even Clarke's *Challenge of the Sea* speculates that the sea might hold the traces of "ancient explorers" from other worlds, which would "at last" grant "proof that our earth is not the only home of intelligence, and we will know that if we once had visitors from space—they may one day come again," and so manages to turn the abyss into a portal to space.) There was no school of Modernist art, perhaps because Surrealism couldn't imagine creatures as weird as those being discovered and Abstract Expressionism couldn't cope with the oblivion of genuine darkness. Without a geopolitical race to the bottom, the deep oceans that had helped launch the age faded to black.[10]

Yet the high ground for military rivalry was not thermonuclear bombs in orbit but nuclear-armed submarines. The search for life that drove the post-Apollo planetary program, particularly to Mars, found nothing, and might discover something only at immense cost and in token fossils. The exploration of the oceans discovered startling realms of previously unknown life, from the largest ecosystems on Earth to exotic niches powered by chemical rather than photosynthetic metabolism. Likewise, the long-anticipated revolution in geosciences came not from other planets but from Earth's oceans. The *Alvin* was reusable and rebuilt in ways that made the space shuttle look like a white elephant. Visiting black smokers along the East Pacific Rise and plucking rocks from the Mid-Atlantic Rift had revolutionized geology more than Moon rocks brought back by Apollo astronauts. And the oceans challenged humanity with tangible threats from rising seas, shrinking reefs, pollution, and overfishing; the only equivalent from space was the possibility of an asteroidal impact. If the key to the Earth's past lay in the ocean, so did its future.[11]

Of the three grand realms for Third-Age discovery, the oceans were the most complex, the most integrated with Earth, and both the most familiar and the weirdest. They generated by an order of magnitude the largest number of expeditions, amassed by far the vastest accumulation of data, found

the most wonders and marvels, and reformed most profoundly the understanding of the one planet humans inhabit. They would sustain the Third Age
while Antarctica slid into normal science and space sputtered and stuttered,
burdened by the search for worlds to colonize and for extraterrestrial life and
intelligence. Paradoxically, "the eternal darkness," as Robert Ballard called it,
would flash with insight as robots searched like tiny needles through unfathomable haystacks.[12]

6

Modern Exploration, Modernist Paradox

Curiosity is moving on. . . . The rover took some time on Jan. 15 to snap
the 57 images needed to generate a fresh selfie.[1]
—C/NET (JANUARY 29, 2019)

When Robert Scott reached the South Pole on January 18, 1912, he wrote into his diary, "Great God! this is an awful place." The scene held nothing but ice and air (and the disheartening token left by Roald Amundsen, who had reached the pole a month earlier on December 14). There was no pole to touch, no rock face to inscribe, no edged border to define the passage, no waterfalls, no pass, no peak, no gorge, no font, no delta. There was nothing to distinguish the scene from the hundreds of miles of ice sheet surrounding it, a continent of ice as wide as Australia, the polar plateau broken only by the tiny wind-carved grooves on its surface. The pole held no tradition of indigenous lore. There was no one to observe, no one to record. The five men took a photo of themselves.

On November 29, 1929, Richard Byrd flew a Ford trimotor aircraft over the South Pole, dropping a small American flag wrapped in a stone. The transition to a new era of discovery—mechanized, sensed remotely—was underway. When Byrd wrote later of what in earlier times could have seemed an epic moment, he noted that there was nothing distinctive about the polar plateau. "The Pole lay in the center of a limit-less plain." It was impossible to know exactly where it was while flying over it. No artfully shaped photograph of the event was possible. He and his crew could photograph themselves in the plane, not on the ice. Byrd could not, as Amundsen and Scott had, put himself into the picture at the site.[2]

As with so many things, Antarctica shows the transition from a Second to a Third Age of discovery. The framing perspective of the Enlightenment was vanishing amid whiteout, vacuum, and eternal darkness. The quests and questions that motivated the Second Age were passing. The means for travel took exploration to places that, if dense with data resisted the inherited intellectual methods for processing them and, if empty, defied efforts to invest them with meaning. Encounters turned inward. The omniscient narrator yielded to an often unreliable self. Dialogues with Others turned into soliloquies. Anthropologists studied explorers rather than the explored. The picture of a world out there imploded into a selfie.

Images of the explorer at his destination or heroically pursuing a quest were a staple of exploration art. But the explorer was seen by those he encountered or by an omniscient artist rendering the scene from a vantage point apart from those depicted. In the Third Age that becomes harder to do, and at some point, perhaps on the Ice, it slipped into something no longer possible or necessary, or at odds with prevailing aesthetics and metaphysics.

Elite culture was stutter-stepping into Modernism, a syndrome aptly characterized by its axiomatic paradoxes, all of which seem to turn on the problem of placing the self within its setting. If the Enlightenment's means no longer seemed to work, perhaps Modernism's would. The self would have to position itself. After *Curiosity*, the American robot destined to roam Mars, landed and unfolded its instruments, it took a photo of itself.

The new technologies put explorers into places that resisted the older interpretive contexts. Those sites challenged painting to render scenes that defied perspective, forced literature to reconsider narration, turned anthropology back on itself, redefined encounter, provoked law and politics to imagine new orders, and compelled natural science to reorganize. At first Third-Age explorers and their sponsors tried to wrestle the new lands and discoveries into old categories, as though Humboldt had landed on the Moon, or Lewis and Clark were traversing Mars, or the *Victoria* was circumnavigating the solar system. But ideas balked. The terranes of the Third Age, as other ages had, needed their own intellectual syndrome. The paradox is that such a syndrome was growing alongside it, but it was a syndrome that had little use for geographic discovery. Modernism had other ambitions. The hardware that made possible the Third Age had leaped beyond the cultural software to run it.

The misalignment resulted, partly, because so much that had defined the Second Age was missing. The intellectual scaffolding could no longer continue to build because the bricks and mortar were missing. Again, Antarctica was the pivot. Its heroic age had sagas as magnificent as any in exploration history, but it could do little other than tell those stories. Everything that had sustained the Second Age was gone. No natives. Few rocks. No terrestrial life away from shorelines. Little to collect. No narratives of social life or cross-cultural interaction. No encounters. There was only the soliloquy and the solo trek. Its tales are stories of survival, not of discovery, or of discovery as personal enlightenment. It was hard to find a new world in which it might be possible to locate even a nowhere such as housed Utopia. Like Byrd's Advance Base, exploration seemed to move beyond the realm of Enlightenment reason.

The trend worsened as explorers probed into space and abyss. At least on the Ice it was possible to breathe and move: the sledging journey held its central place in practice and imagination. In the deep seas and interplanetary space, people could travel only within special habitats that prevented direct contact by human senses other than sight; everything was filtered through prostheses and instruments. Eventually logic and cost argued to remove people altogether in favor of robots. They could do less than people, but they were evolving more rapidly than human explorers, and if instruments were the point of contact, they could do better. As vehicles for colonization, robots stumbled; as instruments of exploration and science, they triumphed. The hybrid human futurists wanted for space travel was evolving, though not as engineered genes or cyborgs but as "exploring systems" in which humans and robots co-produced discovery.[3]

Meanwhile, the Enlightenment was itself undergoing an interpretive crisis that morphed into a metaphysical one. Its informing model, modern science, experienced serial revolutions that had nothing to do with exploration, that came from labs, cyclotrons, and telescopes, not treks. Science's paradigm, physics, was reconstructing its foundations on the very small, the very large, and the very distant, on relativity and quantum mechanics, on statistical mechanics and uncertainty principles. Biology looked to units of inheritance, to genes and gene pools, to new definitions of species not derived from collecting on remote islands and opaque rainforests, and to a modern synthesis of evolution that did not emanate from the insights of far-traveling naturalists. Exploration had less to offer modern science, and modern science less to interest exploration. No scientist would be elected to a national academy on the basis of first-order reconnaissance. No artist would achieve critical and commercial acclaim for painting the terranes of the Third Age. There was no Nobel Prize for explorers.

Those scholarships most closely tied to Second-Age discovery faded into moribundity. Geology took to the library as much as to mountains, with discourses over the origin of granite or orogeny that resembled text-based humanities more than the data-overwhelmed vibrancy in which the discipline had been defined and by which it had inventoried the Earth. Planetary astronomy was a backwater compared with a new cosmologies and astrophysics populated with quasars, pulsars, black holes, and red-shift expansion. Anthropology ran out of new peoples to discover, and turned inward to examine its own society, looking more like sociologists with a heavy baggage train or journalists gone native. The arts moved in lockstep, more interested in the foundations of art than art as a means to unveil the foundations of the exterior world. Painting mostly shunned representation for abstraction, visions of the world rendered without the mathematically framed perspective that had informed Western painting since the early Renaissance. And literature? The Modernist turn abandoned such conventions as omniscient narrators and chronological narrative. The personal narrative meant an interior journey of the artist, new methods applied to understand the nature of writing itself, rather than the journal of a quest for a lost city, the font of a great river, or the rumored Mountains of the Moon. The high culture of Modernism left the apparatus of Enlightenment exploration on the wharf.

The great convergence of art, science, and literature by which Second-Age explorers had discovered or rediscovered the Earth and its peoples broke down, or rather simply ran beyond its supply line. There was no way to live off the land in Antarctica: you lived off whatever you brought with you. That proved as true intellectually as physically, and the farther you went, the less there was left. Those worlds of the Third Age threatened to go beyond not just the systems but the instruments of the Enlightenment. Its sciences had to extend human senses beyond what could be seen, felt, heard, or tasted. Its voyages sailed beyond what humans, without special habitats, could endure. Its literature celebrated states beyond reason, or of outright fantasy.

Literature or history—great narrative depends on character and conflict. But in Third-Age terranes, the parameters are pared to a minimum. In real settings, the conflict is between a protagonist and an abiotic nature, a purely physical contest, or an interior struggle with the self, or else it involves anthropomorphizing robots. Unsurprisingly, science fiction (and technological romance) flourish, not because the scene requires novel science, but because scientific fiction can imagine more compelling conflicts by introducing alien beings or metamorphosing robots into sentient surrogates. Popular fiction found ways to continue the old tropes by conjuring up creatures, mysterious messages, or time travel to periods rife with a usable future. The alternative was to turn inward, which is what elite culture tended to do, examining literature as literature and writing metafiction rather than realist depictions of material worlds, however strange and novel.

In places where there is so little, little can happen. Little matter means little space; few events mean little time. In the real world spacecraft could take years zooming frictionlessly through an unfathomable vacuum; astronauts could cavort on a Moon empty of anything but rock and powder, while robot rovers could trek for months on Mars in search of a rock that might show evidence of past water or even more elusively of a fossil; and submersibles passing through the Earth's abysses and trenches was a voyage into cold blackness, interrupted by an errant jellyfish and stray fangtooth. Even a walk across the Sahara offered sand, rock, sky, sun and stars, mountains and dunes, and the occasional tribesman, either as guide or threat. A flight across the solar system offered only starry sky and occasional flybys of planets spaced by distances measured in years; a plunge to the depths, only an implacable darkness broken by errant bioluminescent creatures and the stray black smoker. That the *Alvin* is best known for visiting the wreck of the *Titanic*, should surprise no one because here it connects with a celebrated and much mulled over human story. Robots as prostheses and projections allow for anthropomorphism. But the twin Mars rovers, *Spirit* and *Opportunity*, don't fit the narrative structure of Lewis and Clark. *Curiosity* isn't Humboldt on Mars. *Pioneer 11* isn't Columbus in the Antilles, nor *Voyager 2*, a Magellan across the solar system.

At its core the critical distinction is the encounter. It was no longer directly between people and people, but as often as not between uninhabitable landscapes and robots. Boies Penrose memorably wrote of the conquistadores that they "embodied much of the best and much of the worst of which the human soul is capable. Their courage was peerless, their cruelty revolting; their endurance was heroic, their lust for riches despicable; their devotion to their leaders was often the personification of fidelity, but the treachery of the leaders to one another was often beneath contempt." "Truly," he concluded, "the conquistadores were men of superlative extremes." However much they might embody the values of their makers, no Mariner spacecraft or Mars rover could exhibit such behavior, conflicts were the result of hardware malfunctions and software glitches, and certainly no rover had a soul. They might journey to the limits of geography; they would not explore the limits of human character. Their stories demanded a different style. They were as distinct from the letters and personal narratives of captains from the 16th century as the Modernist novel is from a late medieval Romance. To shoehorn one into the other verges, as Kurt Vonnegut realized, on parody.[4]

Popular fictions and films were filled with instantaneous travel across boundless light years, intelligent beings, and universal translators that allowed for seamless communication. In its origins *Star Trek* was the voyage of the *Beagle* outfitted with warp drive. If the USS *Enterprise* had to spend decades proceeding to its next destination, there to measure incipient black holes, icy comets, or lifeless planets, there would be no drama, no narrative, no story. Instead, imaginary worlds must be populated with something that resembles Second-Age peoples and marvels. Or perhaps the flaw is with the form. The times might call for the literary equivalent of the selfie, not a genre likely to command popular enthusiasm.

Consider the curious case of the era's grand gesture, the Voyager mission. In many ways the story resembles a quest narrative, but without the obligatory return. Neither of the Voyager twins will reappear on Earth to inform its society and be resocialized. (The original *Star Trek* movie is premised on a future Voyager that does return, to mutual shock.) Rather, Voyager took Earth with it in the form of a golden record full of images and voices, a family-of-man survey of Earth, nominally intended to introduce humans to aliens should such an encounter happen, but actually a message to ourselves about who we would like to think we are. The Voyagers have passed through the solar system carrying a selfie of home.

Even more revealing is painting. The Modernist revolution in art clanked along in loose linkage with that in physics. Cubism, Fauvism, Dadaism, Constructivism, and other avant-gardes ran on a parallel track to quantum mechanics and relativity. That historical marker for Modernism, the 1912 Armory Show, comes midpoint between the special and general theories of relativity. Collectively the newcomers wrenched apart and reassembled the pieces of painting—color separated from line, line from object, object from perspective. They changed how the world was rendered and the role of those doing the rendering. It was not simply a change of topic, like shifting from historical scenes to landscapes, but a reimagining of how to put a three-dimensional world onto a two-dimensional canvas. Yet the loss of perspective is ideally suited to scenes like the Antarctic ice sheet, the empty realm of space, and the dark abyss.

The realms of the Third Age have slapped together hyperrealism and abstraction. There is little to filter the external world, which lends artifacts of human presence a heightened sharpness; but with scant context, they melt into abstraction. Many of the creatures found in the abyss look like emigres from Surrealism. The *Viking 1* photograph of a sunset on Mars looks uncannily like a Georgia O'Keefe sunrise in New Mexico. The liminal ice sheets of Antarctica belong with the flat horizons of Mark Rothko, or better, Robert Rauschenberg's *Erased de Kooning Drawing*, in which he removed an original work and left a blank sheet. The false-colored rings of Saturn resemble the flat, striped canvases of Barnett Newman. Pastiches of streaming galaxies echo the tangled lines of Jackson Pollock. The aura rings of an eclipse look like the blast paintings of Josef Albers. Swirling clouds on Jupiter resembled abstract blotches of a Helen Frankenthaler painting such as *Mountains and Sea* or a Morgan Russell *Synchromy*. The layered blackness of the abyss recalls the themed blacks of Ad Reinhardt; the subtle whites of an ice sheet, the lumpy whites of Kazimir Malevich or Mark Rothko. A portrait of a Mariner spacecraft looks like Marcel Duchamp's *Large Glass (The Bride Stripped Bare by Her Bachelors, Even)*. Yet this was not an art to represent new worlds or one keen to send painters on exploring expeditions. Its subject was art; it was designed to examine the nature of art, not represent new natures.

So while Modernism and its sometimes surly, oft-flaky offspring, Postmodernism, were irrelevant for the terranes of the Second Age, they could characterize with uncanny fidelity the stripped-to-the-minimum conditions of much of the Third Age. The Modernist palette provided ideal instruments for quests to places that lacked people; that, for many, lacked life; and that, for all, lacked traditional frames of meaning. The exploration equivalent of art for art's sake was exploration for exploration's sake—the enterprise stripped down to its purest essence, conducted not just for science or riches too remote to think about but to continue exploring. Of all Modernism's paradoxes, the relevance of Modernist art and the absence of Modernist artists from exploration is one of the most curious. It points to a disconnect between the hardware that could take expeditions to uninhabitable realms and the software by which to endow them with cultural significance.

7

Voyager Traverses the Solar System

You only explore the solar system for the first time once. Voyager did that."[1]

—LARRY SODERBLOM, GEOLOGIST ON THE VOYAGER MISSION

There has never been an expedition like the Voyager mission, and likely there will never be another to rival it. The Voyager twins were launched on August 20 and September 5, 1977, and as 2020 ended they were both still functioning, both beyond the heliopause, both in interstellar space, both the farthest objects launched from Earth. If the space age was announced with Sputnik in 1957, the 43 years of the Voyager mission have claimed two-thirds of the modern era of space exploration. By the time they expire, they will have traversed not only the solar system but 70 percent of the space age. In their longevity, in the magnitude of their discoveries, in the repeated shock of novelties unveiled, in the brazenness of their conception and execution, in the almost literary quality of their narrative frame, they have the trappings of a hero quest. They are the grand gesture of the Third Great Age of Discovery. Nothing else comes close.[2]

The context was an alignment of planets and politics. The planetary dynamics made it possible, in principle, for a spacecraft to visit the outer gaseous planets, one after another, each planetary fling adding critical velocity, in what was quickly dubbed the Grand Tour. As NASA pondered possible missions, it was realized that the next alignment would happen in 1977. The last time it occurred Thomas Jefferson was president, three years away from sending Lewis and Clark across the continent. This time NASA proposed to send two robotic spacecraft across the solar system.

The politics was less predictable. It was the golden age of planetary exploration, and NASA had just sent two Viking spacecraft to Mars to coincide with the American bicentennial. Voyager would carry the flame to the outer planets. But NASA's glory years were rapidly winding down. The Grand Tour was cancelled, then later revived as a mission to Jupiter and Saturn, with the prospect, buried in its hardware and calculations, of perhaps, if the spacecraft completed their tasks at Saturn, of sending *Voyager 2* to Uranus. The technical challenges, moreover, were formidable; the Voyagers' computing power was less than a digital watch today. At launch *Voyager 1* came within seconds of failing to reach its required orbit. During its early cruise phase, *Voyager 2* lost its primary antenna, and completed its trek across the solar system with a backup.

Voyager 1 succeeded spectacularly at Jupiter and Saturn. That freed *Voyager 2* to tweak its trajectory around Saturn in such a way that it could chase Uranus. After Uranus it wheeled its path toward Neptune. *Voyager 1's* flight over Titan took it out of the plane of the ecliptic and toward the stars. *Voyager 2's* flight over Triton did likewise. On March 20, 2013, *Voyager 1* crossed the outer sheath of the heliosphere and entered interstellar space. *Voyager 2* followed in late 2018. They are now moving at 38,000 miles per hour and 34,000 miles per hour, respectively, at a distance of 139 and 115 astronomical units from Earth. They continue to transmit information. By shutting down all but the most essential functions, they have enough power to last another decade.

The returns have been spectacular. The Voyagers were the second spacecraft to visit Jupiter and Saturn; but they were the first to do so with a full panoply of instruments and to survey the giant planets' moons, and the first (and only) to visit Uranus and Neptune. They are the first to enter the heliopause, and *Voyager 1* was the first to exit and pass into interstellar space. They took the first photos of the Earth and Moon together; *Voyager 1* took the first "family album" of the solar system, turning back on February 14, 1990, to capture six planets, including the "pale blue dot" of Earth. They discovered 26 previously unknown moons. They found active volcanoes on Io, warmed by the gravitational tides of Jupiter, and they unveiled geysers on Triton, one of the coldest objects in the solar system. They revealed that the outer planets hold a menagerie of worlds far more active and varied than anyone had imagined. They defined the blustery border of bow shock, where the solar and interstellar winds churn in filmy swirls against each other. They did what you can only do for the first time once. Traversing the solar system is the closest modern equivalent to the *Victoria* circumnavigating the Earth.

The Voyagers' instrumental array was matched by a cultural payload. Their most famous package was a golden record with greetings, songs, and images from Earth—a spaceborne Family of Man. They were headed to the stars; they needed calling cards. The odds that an alien Other might find them and decode their message were infinitesimally tiny (even astronomically tiny, cubed), but they were read by millions of Earthlings. They testified, as all grand gestures do, to what their creators believed was best about themselves. Equally, their classic images became cultural memes. Paradoxically, while celebrated for zooming outward to undiscovered worlds, their best images all looked back. That's how volcanoes were found on Io, how Saturn's rings glowed, how Voyager 1 created the first image of the Earth and its Moon and the first view of the planets from the outside of the solar system looking in. Even as Voyager carried discovery into the future, it looked back to an exploration past. Along with the sights and sounds of Earth, it had for its cultural ballast Earth history.

Part of the Voyagers' appeal is that like Humboldt they were not a scout for commerce or colonies (they bypassed both the Moon and Mars). Although an overtly American enterprise, they carried messages in 55 languages; they were as empty of nationalist geopolitics as possible, as close to disinterested science and exploration as imaginable. They carried neither sword nor cross, nor even speculative platting for a mine, trading factory, or railroad. They were simply explorers.

The Voyager twins began as machines crafted to do science. As they pass through the veil of the heliosphere, however, they are assuming the quality of a quest because more than data, more than images, more than stories of broken hardware and software glitches and ingenious workarounds, more than sheer endurance, the Voyagers have been a journey. Their cultural payload is what distinguishes the mission from adventuring; their trek is what has distinguished it from space science. What Robert Frost once said of a poem, that like a block of ice on a hot stove, it should ride on its own melt, can be said of the Voyager story. Its narrative rides on its own journey, even if, to obtain closure, it contains a self-referential loop.

8

New Realms, New Regimes

The process has already started and will lead to a competitive scramble
for sovereign rights over the land underlying the world's seas and oceans,
surpassing in magnitude and in its implication last century's colonial
scramble for territory in Asia and Africa. The consequences will be very
grave.[1]

—ARVID PARDO, SPEECH TO UNITED NATIONS, NOVEMBER 1, 1967

The discovery of previously unvisited places presented not only an intellectual challenge but an institutional one. It was not enough to locate newly discovered places and peoples into a schema for understanding: they also required protocols for behaving. They compelled reforms of politics, law, and governance—and offered new opportunities for millenarianism. And so it proved with the Third Age. The relations among discoverers and between them and the discovered had to be unpacked and reassembled.

But this time was different. The corroding contact between those discovering and those discovered was gone. There were no indigenes in Antarctica, on the ocean floor, the Moon or Mars. There would be no clash of arms, no forced conversions, no treaties, no trade, no paternal oversight. Much of what had tainted exploration in the previous centuries no longer applied. What took its place, in muted ways, was a worry over biotas. Guidelines were constructed to prevent unwanted contamination of native ecosystems or the return of alien species to Earth. Legal (if not political) rights were extended to un-Earthly forms of life (and in science fiction, over sentient robots). Mostly this remained in the realm of the highly hypothetical—routine fodder for popular fiction and film, but nothing that would push legal regimes or courts beyond their elastic limits.

The relationships within and between the exploring nations were another matter. The Third Age, like its predecessors, was charged by competition that needed regulation. Antarctica, the oceans, and near-Earth space were accessible to commercialization; all three realms were open to militarization; and what the exploring nations did would affect the globe. The newly accessible geographies needed a suitable legal and political regime.

The emerging order built on the inherited one. Over the centuries a corpus of international law had elaborated on what constituted discovery and on what basis a nation might claim rights to discoveries. A protocol for discovery evolved. When islands and coastlines were the prime realms, first sighting was deemed sufficient until other explorers demanded a more tangible act, a landing along with a formal declaration and a monument of some kind. Portugal and Spain had no sooner begun voyages, however, than they quarreled over the real and potential spoils. The treaties of Alcáçovas, Tortesillas, and Zaragoza divided the globe between them. Competitors protested. They raided coasts, abducted treasure ships, and published legal rejoinders like Grotius's *Law of the Sea*.

More complicated was the penetration of continents. Trading posts and colonies escalated the requirements into "effective settlement," and then into various concepts of "hinterlands," seeking to establish just how far a presence might justifiably extend into the backcountry, a judgment that had to factor in the character of the land and its inhabitants (if any). The Berlin Conference exploited the concept of hinterland to guide the partition of Africa. In the early 20th century the practice even extended, however improbably, to Antarctica. These issues were still under legal definition as late as 1933, when the international court at The Hague ruled that Norway could not claim eastern Greenland, since Denmark's two settlements on the south and west were sufficient to extend its rule over the uninhabited ice sheet.[2]

The project could take on a surreal quality when extended to the realms of the Third Age. What rights of discovery might mean on an abyssal plain, an ice sheet, or a lifeless moon was academic until someone arrived and made claims. The critical concepts were the high seas and the terra nullius. These were the places that in earlier eras were empty of human inhabitants, or in some cases cynically treated as though they were empty. In the Third Age they were unambiguously empty, apart from the presence of other explorers, their satellites, and surrogate rovers.

Predictably, the process began in Antarctica. During the early 20th century, seven nations had laid claims—three of which overlapped (Britain, Argentina, and Chile each insisted on the Antarctica Peninsula). One region of West Antarctica was formally unclaimed but generally recognized as the American sector; in 1924 the United States declared that it would neither advance any claims nor acknowledge those of others. The claims were mostly meaningless since no one apart from the claimant recognized them. They could, however, destabilize and lead to a scramble that no one wanted. Then IGY changed the dynamic. For its 18-month duration all participants agreed to treat ice as high seas, open to all. The United States established a major base on Ross Island and at the South Pole; the USSR, which had not visited Antarctica since Bellingshausen in 1820, plunked bases down in every claimant's territory. The Cold War had come to the Ice. Since neither rival had vital interests, and certainly didn't intend to fight over the continent, any political tensions were frozen. There were no critical resources at stake. Antarctica was far removed from iron and bamboo curtains. Collaborative science provided cover for imposing a cold peace.

The success of that experiment encouraged the search for a permanent replacement. There was some urgency since several nations announced intentions to make temporary bases permanent in what looked like settlement by stealth, or what might count as facsimiles for the military bases and trading factories of the past. In 1959, at the invitation of the United States, interested nations convened in Washington, D.C., and emerged with an Antarctic Treaty destined to enter into force in 1961. The treaty established a governance regime, what has come to be called the Antarctic Treaty System (ATS). It disarmed territorial politics by refusing either to accept or to deny existing claims: Argentina and New Zealand and the others could continue to declare sovereignty for domestic consumption, while everyone else ignored their proclamations. It was another Modernist paradox, the political and legal equivalent of Bohr's principle of complementarity in which an electron might be both wave and particle depending on circumstances.[3]

The treaty also distinguished between two categories of participating nations. Consultative nations established bases and programs—they had some tangible investment in what happened. Signatory nations simply agreed to the ATS by signing. The consultative nations formed an approximate, de facto parliament. The ATS established common codes: Antarctica would be demilitarized, would not allow nuclear testing or accept nuclear waste, would resolve disputes through the International Court of Justice. Other protocols have been added to cover a mineral regime, marine mammals, and environmental concerns. Over the years new countries like China, Brazil, and India have acquired consultative standing, defusing potential conflicts by absorbing discussions within the ATS. Not least, the treaty provided for a review and reconsideration after 30 years, but no nation has demanded one. All parties appreciate that the situation is metastable, and no one believes that if the treaty is dissolved, it could be reconstituted into anything like its current form. The Antarctic Treaty finessed national sovereignty into something more symbolic than operational.

After Sputnik in 1957 and Explorer in 1958 were launched, concerned parties fretted over nuclear arms in orbit, or more generally the militarization of space, especially with weapons of mass destruction. Though the first satellites had been launched under the aegis of IGY, their military origin placed them squarely within the Cold War and so frustrated negotiations.

Addressing the UN in 1960, President Dwight Eisenhower urged that the principles enunciated in the Antarctic Treaty be adapted and extended over the solar system. Near-Earth space introduced issues not relevant in remote Antarctica: satellites, perhaps militarized, could orbit over sovereign territories. In 1963 the Limited Test Ban Treaty removed the question of orbiting nuclear weapons. Other issues were resolved, with one objection after another overcome, mostly along lines analogous to those for Antarctica. Meanwhile, the space race added urgency. Weather satellites, communication satellites, and surveillance satellites were multiplying; spacecraft passed by Venus and

Mars; other craft flew around, into, and onto the Moon. There was good reason to believe that, however inappropriately, the classic contests sparked by exploration in the past would extend beyond Earth. In 1966 the General Assembly recommended approval of what became known as the Outer Space Treaty. In 1967 it was opened for signature and entered into force.[4]

The treaty declared that the "exploration and use of outer space" should be conducted "for the benefit of all peoples," that such realms "shall be the province of all mankind," that space should not become the arena for "a new form of colonial competition." That phrasing looked generally to the useful ambiguities of the Antarctic Treaty and to the heritage of the open seas. Among its provisions are prohibitions against placing into orbit nuclear arms or other weapons of mass destruction; against claiming sovereignty over the Moon or other celestial bodies; against contaminating extraterrestrial objects; against installations closed to others. Nations are responsible for their activities in space, including damage caused by falling debris. They are bound to assist astronauts in distress.

Antarctica had a history of exploration and territorial claims, which the ATS had to navigate around. Space did not, as yet. When Eisenhower announced his proposal, three years after Sputnik, no one had set foot or planted a token on the Moon or any other celestial body. That changed three years after the Outer Space Treaty was opened for signatures. The treaty barely beat boots (and bots) on the Moon.

Far more complicated was the ocean. Earth's seas had been the subject of extensive international treaties and law since the Great Voyages. Between 1479 and 1529 Portugal and Spain negotiated a suite of treaties that partitioned the discovered new worlds between them. The first divided Morocco and the Atlantic isles—Spain got the Canaries, and Portugal the rest. The most critical determinations involved the primary realm of the First Age, the Ocean Sea. The Iberians naturally wanted to rule the seas as they did isles and the discovered lands. In practice, however, they were unable to enforce such an edict, and the more nations that joined the contest, the more difficult it was for any one to assert unquestioned ownership.

Instead of a right to a mare clausum, or closed sea, challengers proposed a mare liberum, or open sea. In 1609 the Dutch jurist Hugo Grotius codified those arguments into a book of that title, establishing a body of customs and prescriptions that evolved into a generally recognized Law of the Sea. It was not an easy sell. That the Dutch were, at the time, keen competitors demonstrated the working alliance of politics, commerce, and jurisprudence. Self-interest figured as fully as idealism. Grotius was himself equally engaged in justifying various Dutch intrusions, such as seizing everything from vessels to ports, particularly against the Portuguese and English. Nations had a right to their coastal seas out to three miles (roughly, the range of a cannonball), but the idea that the high seas should be freely open to all became an enduring legacy.

The deep oceans were irrelevant. There were no resources to harvest, no passages to thread, no naval threat, no access of any kind. The prospects were so remote that they went beyond the hypothetical to the hallucinatory. That changed as submarines altered naval power, and the Cold War went deep as well as long, and seafloor mining became technologically plausible. The ocean floor was accessible, knowledge about bathymetry vital, and issues of governance critical. The world sea was collecting problems as it did garbage; fishing and drilling on continental shelves made access desirable; military traffic brought them into the realm of national security; and schemes for largescale commercialization and even colonization abounded. The mare liberum threatened to become a submarine mare clausum. The deep oceans needed a legal regime.

The Antarctic Treaty was again an obvious model, but the ocean had millennia of human use and was not, as ice and space were, a blank slate. It was not obvious how to segregate the surface seas from the ocean depths. Legacy would have to balance with discovery. The process commenced at the end of IGY with a 1958 UN Conference on the Law of the Sea, which resulted in four updates. For the seafloor two concepts competed. One regarded it as the bottom of the high seas, and hence open to all; the other, as adjacent land to continents, which is to say, the hinterland of coastal nations. The first favored all peoples equally. The second played to those with coasts and continental shelves and those who had the power to exploit whatever resources the deep might hold—manganese nodules seem to function as the alluring nuggets of the abyss. The first, extending traditional notions for land, regarded the ocean floor as a terra nullius, a land of no one. The second recognized the novelty of the setting and the presence of institutions like the United Nations, and argued for a *terra communis*, a land of everyone.

Negotiations began through the UN in 1967, just as the Outer Space Treaty went active. Malta's ambassador, Arvid Pardo, picking a phrase from the World Peace Through Law Conference, proposed that the deep seas and ocean floor should be the "common heritage of all mankind" and that any commercial use be overseen by an organ of the UN. They would be governed by all and ruled by none. Alarmed by the prospects for militarization and inspired by possibilities that expropriation by military and mining interests had not yet matured, Pardo urged that the nations of the world "examine whether it might not be wise to establish some form of international jurisdiction and control over the sea-bed and the ocean floor underlying the seas beyond the limits of present national jurisdiction, before events take an irreversible course."[5]

The existing Law of the Sea was complicated in ways that legal regimes for Antarctica and Outer Space were not, and the "common heritage" principle was a potentially destabilizing concept. Discussions continued until 1982 when a new Law of the Sea Convention was open for signature. The convention extended the territorial seas from three to twelve miles, established an exclusive economic zone of two hundred miles from coasts, and otherwise granted the deep oceans the same freedoms as the surface seas. With one exception: adopting the "common heritage" principle, it established an International Seabed Authority, under the auspices of the UN, to oversee exploitation of the ocean floor. That principle the United States refused to recognize, and accordingly it has yet to sign the convention. (The "common heritage" concept underwrote a second treaty for space, one nurtured on the example of the Law of the Sea rather than the Antarctic Treaty. The Moon Treaty, completed in 1979, entered into force in 1984 for the thirteen nations that have acceded to it.)[6]

Complicating the picture are the inevitable ideologues, free riders, and Third-Age equivalent to privateers and freebooters—the Cecil Rhodeses and Leopold IIs—who want untrammeled access to new lands and unfettered rights to claim them as private fiefdoms. They speculate about creating stateless islands on the high seas, settlements on Mars, or mining camps in the asteroid belt. Seasteading and lunar colonies are the new island utopias, conveniently beyond the reach of government, societal obligations, and taxes. But they are less the fantasies of intellectuals than of the 21st century's super rich. They are not model societies but libertarian refugia, with a justifying patina of futurism. Their intellectual genealogy traces to Ayn Rand rather than Thomas More.

The social costs and potential governmental conflicts are real. Antarctica already suffers from unregulated tourism; the oceans are becoming fish deserts and gyres of garbage; near-Earth space is cluttered with debris from commerce and the clamber from well-heeled tourists, and its near militarization threatens to unhinge existing treaties. The prospect of a scramble-for-Africa on the Moon or Mars should be enough to frighten even the most fanatic partisan. Private profits and schemes can have social costs. Stateless entrepreneurs roaming the depths and across space might easily have the destabilizing presence of stateless terrorism.

All this the Third Age has set into motion. The aftershocks will not end soon.

9

Before and After

The further people have traveled from the home planet, and the more the universe around it has expanded, the more precious and unique the Earth has appeared.[1]

—ROBERT POOLE, *EARTHRISE*

In 1962—a year after the Antarctic Treaty entered into force, the year *Mariner 2* made the first planetary flyby (past Venus), two years before the DSV *Alvin* launched—Thomas Kuhn published *The Structure of Scientific Revolutions*; a second, revised edition appeared in 1970. In its pages Kuhn argued that scientific understanding toggled between periods of sequential "normal science," organized around what he termed an intellectual *paradigm*, punctuated by moments of abrupt change, what he termed a *revolution* in which a new paradigm replaced the old. The concept was provocative, the thesis a veritable revolution almost in itself, but whatever its merit, the timing of its appearance coincided neatly with a suite of scientific revolutions catalyzed by the Third Age. In fact, for some partisans, Kuhn's meditation became a manifesto.

As with earlier ages of discovery, the Third poured forth a cornucopia of fresh knowledge, devised tools by which to reexamine even long-known places and inherited knowledge, and it rebuilt institutions. It was a question of information mass and density, not just of kind; and a velocity and cascade of discovery, not just novel facts, however startling. The volume of data overwhelmed legacy systems, too much and too fast to be absorbed piecemeal. New disciplines proliferated to cope. In short order astronomy morphed into space sciences, geology into earth sciences, and oceanography into marine sciences. New systems of professional societies, advisory committees, and funding evolved in tandem.

It was soon not enough to catalog the new discoveries, creating high-tech equivalents for cabinets of curiosities. With considerable labor the pieces came together to fashion an updated cosmology that reimagined Earth as a planet and repositioned it within the solar system. "The space programme," as Robert Poole pondered, "which was meant to show mankind that its home was only its cradle, ended up showing that its cradle was its only home."[2]

SPACE

Astronomy had the longest history as a science—the name itself dates back to the 13th century. Newton's *Principia Mathematica* had used the known solar system to illustrate his laws of motion and his conception of gravity, and the resulting Newtonian cosmos became the exemplar of Enlightenment. Or as Alexander Pope put it, "Nature and Nature's laws lay hid in night: / God said, Let Newton be! And all was light."

Great shifts had occurred in the early 20th century, propelled by giant telescopes and the ability to measure infrared and X-ray and to engage in spectral analysis, but they focused on quasars, pulsars, distant galaxies, and an expanding universe. The looked beyond—literally looked past—the solar system. What became planetary science did not exist as a separate discipline. Rather, it meant the application of Newtonian mechanics to the study of planets as celestial bodies, with whatever data could be derived from optical telescopes. When the newly created U.S. Air Force funded a summary series of books on planetary astronomy in the early 1950s, Gerard P. Kuiper of the University of Chicago's Yerkes Observatory was the world's sole professional practitioner. When Nobel Laureate Harold Urey delivered the Silliman Lecture in 1951 on "The Planets: Their Origin and Development," he devoted 55 pages to the "terrestrial" (inner) planets and a meager five to the "major [outer] planets and their satellites."[3]

The primary college textbook was Robert Baker's *Astronomy*, first published in 1930 and revised periodically. The seventh edition appeared in 1959, barely capturing some of the results of upper-atmospheric and solar data from IGY. The eighth edition came out in 1964, just on the cusp of revolutions powered by Third Age discoveries. What it knew about planets and moons was what Earth-bound telescopes could detect and what Newtonian mechanics could deduce. In fact, apart from chapters on stellar systems beyond the solar system, the state of knowledge was not much advanced from Newton's day.

The solar system was analyzed by the size, mass, orbits, and rotations of its bodies, and their interaction through gravity. Earth's Moon was mostly about orbits, tides, phases, rotation, and on its surface, albedo, temperature, and the physiography of mountains, maria, and craters. Regarding its origin the two prevailing theories were igneous and impact, both based on analogies to Earth equivalents. Similar analysis applied to the inner planets, though their surfaces were obscured by proximity to the Sun (Mercury), by clouds (Venus), or seasonal changes (Mars). A full page, with photos, was devoted to Martian "canals," concluding that a "difference of opinion persists today," and that "the times of the Martian year when the dark markings change in intensity and color are such as would be expected if the changes are caused by the growth and decline of vegetation." Jupiter was distinguished by its size, its clouds (including the distinctive Great Red Spot), and its 12 satellites, divided between inner and outer moons. Saturn was distinctive for its rings, which take up 4 of the 6.5 pages allotted. Uranus and Neptune claimed 2.5 pages between them, of which half was opaque black-and-white photos. Kuiper lamented the lack of "reciprocity" between geosciences and planetary astronomy; the latter could only marvel at the "incredible richness of the data" the former possessed. In short, not much was known, and controversies were unlikely to be resolved by continuing inquiries of the same sort.[4]

Then exploration blasted off and the data streamed back. IGY required three world centers to hold the rising tide. *Explorer 1* found the Van Allen radiation belts. *Tiros 1* began imaging the dynamics of Earth's atmosphere, and even before the Apollo program was announced, it had photographed all the continents save Antarctica. Lunar Orbiter, Surveyor, Ranger, and Luna spacecraft went to the Moon; *Luna 3* imaged the far side. *Mariner 2* flew by Venus; Mariners 3 and 4 zoomed past Mars. *Pioneer 6* amassed tape recordings, "shipped daily, big 9,600 foot, 17-inch reels"—"truckloads of tapes." Astronauts landed on the Moon and returned rocks. Viking sampled the Martian surface. Voyager toured the outer planets. Writing in 1981, when the Grand Tour had just begun, three prominent geoscientists remarked that their field was "immersed in a planetary information explosion." As *Voyager 2* rushed toward Uranus, JPL estimated the volume of data the mission had so far dispatched to Earth as four trillion bits, enough to "encode over 5,000 complete sets of the *Encyclopedia Britannica*." Then *Galileo* orbited Jupiter; and *Cassini*, Saturn; and *New Horizons* bolted past Pluto and its moons.[5]

What had been summaries of the planets became mere sidebars. What had endured for decades as near-scholastic discourse about Martian canals, lunar maria, and Jovian clouds ceased as hard data overwhelmed existing theories. Planets became miniature solar systems, rich with orbiting moons, each distinctive. Io had volcanoes, Triton geysers, Europa a subice ocean, Titan a methane atmosphere. Mars had ice caps, Mars had water. The surface of Venus was mapped by microwave. The clouds of Jupiter were digitized by *Galileo*. Orbiting telescopes like Hubble revealed a wider universe unfiltered by Earth's atmosphere. Looking back in 1985, Oran Nicks observed that "it is not easy to recapture the extent of our ignorance a quarter-century ago; *everything* we learned was new." Analogies animated theories as scientists struggled to describe first contacts. What had been a page and a grainy photo in a pre-IGY textbook now deserved its own book, updated frequently.[6]

Twenty years after the University of Chicago Press published *The Structure of Scientific Revolutions*, Cambridge University Press published *The New Solar System*, overflowing with color photos, maps, graphs, and data that inflated the data reserves of planetary astronomy as New World silver had the economy of Renaissance Spain. The solar system was bursting with new worlds. Paradigms shifted.

EARTH

No planet felt the impact more than Earth itself. Almost hard-welded to the Second Age, geology found it difficult to overcome its inherited perspectives, and conducted debates with ever more clever parsings and rereadings of existing data. The Second Age had filled libraries and laboratories with reports, books, specimens, data, which geology had used to address the age of the Earth and the organization of geologic time. Yet the science seemed unable to answer the questions that replaced them, why the great features of the planet were where they were and how they interacted. Then, after World War II, the earth sciences underwent the revolution that many of its practitioners yearned for.

Begin with Antarctica. When the heroic age ended, only a fraction of the continent's shoreline had been mapped. Two treks had gone to the South Pole, and one to the south magnetic pole. A handful of other, shorter sledging journeys had been completed. The annual rhythm of the flanking sea ice was known, but again not mapped, or surveyed sufficiently to portray average conditions. (Shackleton thought the Weddell Sea would be mostly open; it was mostly iced over.) It was agreed that Antarctica consisted of a greater (eastern) and lesser (western) part and that they differed in character, but there was no understanding of how they might be related. Nothing under the ice was known.

By the time IGY concluded, and polar exploration was segueing into normal science, the continental coast was mapped, the continent had been traversed and overflown, seismic profiles had sketched the gross geography and bathymetry of the ice sheets, the basics of Antarctic weather were understood, and ice flows were measured. Within another decade earth science had delineated the context of Antarctica within plate tectonics and planetary history. There were tourist overflights, adventure treks, and cruise ships along the peninsula. What had been as little known as a moon of Saturn became a familiar—if uninhabited and hostile—part of Earth.

The revolution that moved continental drift from a crank theory to a foundational doctrine did not need Antarctica. Its critical data came, paradoxically, from the oceans; had geologists continued to collect samples from the continents alone, they might never have understood the mechanism. When in 1912 with lectures and in 1915 with a book, Alfred Wegener, a German meteorologist and geophysicist, announced that the continents had moved into their present positions, the notion met with open skepticism. Then Wegener, a polar explorer, died in Greenland in 1930, and his idea perished with him, as if Second-Age methods could not answer Third-Age questions. There was no plausible mechanism by which continents could push along through the crust. Instead, the idea of a cooling Earth that had wrinkled ("like a drying apple") into mountains and oceans still prevailed. Or it did until research into the deep oceans proposed the necessary mechanism and plate tectonics was born.

"Earth science is ripe for a major scientific revolution," wrote J. Tuzo Wilson in 1963. Wilson had headed the International Union of Geodesy and Geophysics that oversaw the scientific program for IGY, had grasped the global perspective it inspired, and fumed that geology so lagged behind the 20th century's reformations that had swept over discipline after discipline. In his retrospective account of IGY, he fretted that "No one knows with any certainty how the earth behaves, why mountains are uplifted, how continents were formed, or what causes earthquakes. We know some anatomy of the earth, but no real physiology." A year earlier Harry Hess had proposed, based on the new data from the oceans, that the seafloor spread; a year later a Symposium on Continental Drift worried less about the reality of a moving crust than plausible mechanisms behind it. By the end of the 1960s all the data dots were being connected. Before then, in Wilson's words, "all the pegs seemed square and all the holes round. In each case, it was not until it was realized that one had to discard the whole frame of reference and seek another that answers came in a flood." Plate tectonics crystallized. In a way almost calculated to parody Kuhn's concept, a new paradigm suddenly appeared.[7]

The shift was comprehensive. All aspects of earth science now aligned less along the question of Earth's evolution than around problems of its dynamics. Piece after piece of the Earth puzzle—mid-ocean volcanoes, shuffling continents, long faults through places like California, the gathering of mountains along continental coasts—found its correct conceptual location. It's hard to imagine the riddle resolved had geologists continued their inherited methods, no matter how often they pilgrimaged to legacy sites or reviewed the classic literature. Or as Henry Menard notes, prior to the Third Age, "geology was barren of testable ideas." The sought-after answers came from elsewhere, from sites invisible to the Second Age, but places that became prime realms for the Third.[8]

OCEANS

They came primarily from the oceans. What the space program promised, ocean exploration delivered. This seemed historically apt since oceanic exploration had led each of the three ages. In 1876 reflecting on the completion of the *Challenger* expedition, Sir Charles Thomson observed that "strange and beautiful things were brought to us from time to time, which seemed to give us a glimpse of the edge of some unfamiliar world." A century later hints and glimpses had been replaced by dense data sets, labs overflowing with core samples, sedimentary and biological collections from the deep, and remarkable visits. The abyss remained unfamiliar, but it was no longer unknown.[9]

Prior to IGY oceanography's great compendium was *The Oceans: Their Physics, Chemistry, and General Biology*, published in 1942 and written by three professors at Scripps Institution of Oceanography—Harald Sverdrup, Martin Johnson, and Richard Fleming. Like Baker's *Astronomy*, it mostly applied Newtonian physics and standard chemistry to understand the character of seawater and how the mass of the ocean behaved, and it appealed to the biology of the day to explain how life functioned in the waters. It seemed a lot: 1,086 pages in all. Yet it mostly positioned the world's seas within a matrix of existing disciplines. Compared with the area and volume of the ocean, there had been few surveys, only tiny patches of bathymetric mapping, a handful of deep samples, and unsystematic records of fishing. By elaborating carefully on what was known, the authors seemingly filled the great basin of oceanography.

Twenty years later, after postwar expeditions and IGY, much smaller works packed greater punch, exemplified by Harry Hess's 1962 article on the "History of Ocean Basins." Textbooks more resembled communiques from the front lines as data swelled into a tsunami and the ocean turned dynamic, no longer the tomb-like, water-filled bathtub of the planet, but the engine of planetary tectonics and weather; not the dumping ground for what was left after the continents, but the motive source for those continents. *The Oceans* had a postwar edition (1949); then Harald Sverdrup died in 1957, as IGY began. Others took up the revisions, first after IGY (1961) and then during the revolution in earth science (1964, 1965, 1970); and more followed. Oceans went from being a sink to a source.

For a host of reasons the promises of the Third Age were unfolding less in the infinity of space than in the depths of the Earth's seas. The oceans are closer, and cheaper to explore, and however alien to humanity's quotidian existence, the seas are continuous with humanity's world both geographically and historically. The "void" of the deep seas is in reality a tangible medium, a watery matrix and a more plausible nursery of life than microbe-carrying meteors from Mars. Far from being abiotic vats, only capable of supporting life at the surface, the oceans are the dominant habitat on Earth; however hostile to terrestrial life, the sea is an abode for Earthly life and is continuous with it. Increasingly it appears that the rifts, submarine volcanoes, hot vents, and black smokers that parse and stitch the solid floor of the sea are fundamental to a grand geochemical cycling of planetary water, and with the nutrients and minerals essential to a living planet. The Law of the Sea is where geopolitics is being fought. Not least, it is possible to step from land to sea and back again. The shore is a familiar boundary between the two, literally ebbing and flowing with the rhythms of Earthly existence. One can jump into the ocean by stepping into the surf, but into space only by massive rockets and a leap of faith.

Over and again, the most publicly ardent proponents of planetary exploration said they saw their endeavor as part of a larger mission to colonize, and an astonishing number traced their enthusiasm to an adolescent literature of technological romance in which worlds were found and lost in space. While the oceans had their lore, and champions of the sea held fiercely to their distinctive sagas of exploration, that tradition stayed on the surface. The utopians wanted colonies on other worlds, by which they meant worlds that looked like planets. What the deep oceans showed, however, was that exploration was neither confined to space nor defined by it.[10]

The deep seas were, at the dawn of the Third Age, mostly a deep mystery. A century after Maury had synthesized the known ocean in his *Physical Geography of the Sea*, Rachel Carson summarized the state of knowledge in her prize-winning *The Sea Around Us* (1951). Maury's geography appeared two decades before the *Challenger* expedition; Carson's, two decades before the revolution in earth science unleashed by discoveries in the depths.[11]

The Sea Around Us was an extended natural history essay with a literary cabinet of curiosities, each specimen lovingly polished, on such topics as tidepools, the fabulous tides in the Bay of Fundy, islands, the giant squid. Here was oceanography as natural history, history as anecdote, nature as parable; *The Sea Around Us* is the literary counterpart to Sverdrup's *The Ocean.* "Here and there, in a few out-of-the-way places," Carson muses, "the darkness of antiquity still lingers over the surface of the waters." But it is "only in thinking of its third dimension that we can still apply the concept of the Sea of Darkness." In 1951 the depths were unknown in any serious way. Echo sounding was barely underway; seamounts were surprise discoveries; the records of the Challenger expedition were still a baseline. So it was appropriate that, throughout, there was a prophetic aura since Carson believed that Earth's seas, its collective Oceanus, were the grand synthesizer of geology, climate, and life, "the beginning and end," for "all," she insisted, "at last return to the sea." Besides, she intoned, it is "always the unseen that most deeply stirs our imagination."[12]

A decade later Carson republished the book with a revised preface and an appendix of notes, as Tuzo Wilson was publishing his *IGY: The Year of the New Moons*, a memoir of his tenure as president of the International Union of Geodesy and Geophysics during IGY. The "failure to recognize the continuity of the mid-ocean ridge until 1956," he lamented, was a glaring indication of how sclerotic geology had become. "Real progress" depended on a flotilla of war-surplus research vessels with their sparkling instruments—their hydrophones, echo sounders, fathometers, piston corers, cameras, explosion seismometers, magnetometers, dredges, and thermal probes. Quickly, they stripped the veil from the abyss.[13]

The imagination no longer needed to rely on hypotheticals, poetic tropes, mythopoeic incantation, and appeals to "ultimate causes": it could feast on a smorgasbord of hard data. Carson's magical isles had become hundreds of seamounts; her giant squid, whole new biotas; the ecology of sloshing tidepools, the bizarre ecosystem that gathered around vents and black smokers; the majesty of the Fundy tides, the crustal cycling of tectonic plates, plunging under and overriding continents to raise the Andes and the Himalayas. Readers still relish her book for its graceful prose and sensibilities, as they might read Thoreau on the Concord woods.

But the real art of deep-sea discovery was drawn with the bathymetric maps of Marie Tharp and Bruce Heezen. In 1957, as IGY launched, they published the first more or less comprehensive view of the seafloor of the north Atlantic Ocean. In 1977, after the earth science revolution was consolidated, the same year the Voyager mission launched, they expanded that cartography to encompass all the world's seas. (Heezen died that year while on a research submarine near Iceland.) Between the seafloor and the subice terrane of Antarctica, the Third Age mapped 80 percent of the Earth's solid surface for the first time. The newly seen could stir the imagination as deeply as the unseen. There was real poetry in that achievement.

In 1959, a year after IGY and the creation of NASA, Kurt Vonnegut Jr. published a science fiction satire, *The Sirens of Saturn*, in which he mocked the presumptions of the starry eyed. "The monsters between space explorers and their goals were not imaginary, but numerous, hideous, various, and uniformly cataclysmic; the cost of even a small expedition was enough to ruin most nations; and it was a virtual certainty that no expedition could increase the wealth of its sponsors." A decade later *Star Trek* began its endless frontier as a TV show-cum-movie franchise with the promissory prologue "to explore strange new worlds, to seek out new life and new civilizations, to boldly go where no man has gone before."[14]

Neither proved true. The Third Age was neither all novelty and wonder nor all hazard and horror. The exorbitant cost of discovery was absorbed into the budget of the Cold War, and vessels safely landed on the Moon, Venus, and Mars, orbited Jupiter and Saturn, passed through the solar system, and plunged into the deepest canyons of Earth's oceans. The marvels were real, yet different from those anticipated, and light-years from those that powered the technological romances of the era. Imminent colonization looked more and more like a chimera, commercial exploitation still clinging to near-Earth space like starfish in a tide pool. The explorer was most often a robot. The moral drama of the age flickered uncertainly across history like an aurora. Yet the age was as real, as vital, as fascinating as those that preceded it, and the Earth was different for its happening.

10

Looking Back, Looking Ahead

These days there seems to be nowhere left to explore, at least on the land area of the Earth. Victims of their very success, the explorers now pretty much stay home.[1]

—CARL SAGAN, *PALE BLUE DOT* (1994)

How should we understand the Third Great Age of Discovery? Like its most spectacular expression, Voyager, taking photos of the Earth and Moon and of the solar system even as its trajectory pointed beyond the heliosphere, the age looks back as much as ahead, triangulating between past and future. The other Great Ages have their historical frames set: we can compare each to what existed before and what came after. The future of the Third Age has as yet no bookend.

Like its predecessors the Third Age has its special realms of geography, its technologies, its intellectual syndrome, its geopolitical rivalry, and its peculiar moral drama. The Third Age explores ice, space, and abyss. It relies on novel technologies of propulsion; innovative sensors that can record through cloud, ice, and water; and, with computers, radically fresh ways of processing data. It operates within the larger context of a Greater Modernism, an intellectual perspective different in some of its fundamentals from the Renaissance and the Enlightenment. It grapples with the absence of an Other: it is, after a fashion, the culture talking to itself, or the cultural equivalent of echo sounding, refracted through remote regions of the planet and solar system. Its enduring emblem may be *Curiosity*'s multi-imaged selfie on Mars.

It is still young. If we date the First Age from the sacking of Ceuta (1415) and the discovery of Madeira (1420) to the death of William Dampier (1715), the era extended across 300 years. If we begin the Second Age with the Paris Academy's expeditions to Lapland and Ecuador (1735) or with the first voyage of James Cook (1767) and end with Ernest Shackleton's abortive attempt to traverse Antarctica (1916), we have 181 or 149 years, or between 150 and 200 years duration. If we begin the Third Age with Operation Highjump in Antarctica (1946) and use as a marker the year 2020, we have 74 years. The Second Age lasted half as long as the First, and at present the Third is half as long as the Second. What its final length will be is unclear, and will depend on how we define the segue from exploration into science and adventuring; but it will probably be less than its predecessors. As with previous ages, the first astonishing surge of discovery has lasted roughly a generation, perhaps two. The Third Age had an explosive start, jolted to life by the electric charge of the Cold War, but it has settled into the measured pace of a marathon. Still, it likely has a good run yet to make.

Will the Third Age more resemble the First or the Second, or some hybrid of the two? In space and on ice it seems likely to follow the example of the First, with a few complex and costly expeditions to new lands and with outposts—scientific rather than commercial or military—that are not destined to become outright colonies. The oceans are different; far denser in discovery, more elaborate in expeditions, more sensitive to national rivalries for military and commercial advantage. By number and output the deep seas will, over the coming century, define the era. Too many proponents of space exploration look only outward. They need to look inward as well.

Comparing their styles, consider the Canary Islands as a common reference. In the First Age the Canaries were among the first Atlantic islands discovered (or rediscovered) by Europe; they were contested by the primary rivals, Portugal and Spain; and after Spain triumphed, they became its principle point of departure for the New World. Its settlement was a model for New Spain; its great peak at Tenerife, a beacon for mariners and the port of embarkation for an emerging global transit network. In the Second Age the peak was rediscovered by naturalists. "The port of Santa Cruz," observed Humboldt, "is in fact a great caravanserai on the route to America and India. Every traveler who writes his adventures begins by describing Madeira and Tenerife, though the natural history of these islands remains unknown." Humboldt then proceeded to redefine El Pico de Teide as a stele for interpreting the geography of land, a symbol transported to the New World and projected onto Mount Chimborazo. The Third Age finds the Canaries again a port of call, this time for undersea exploration, a search for submarine Teides like black smokers that present a weird inversion of the life zones traced by Humboldt. The FAMOUS expedition to the Mid-Atlantic Rift passed between the Azores and the Canaries, and in March 2007, the RRS *James Cook* left the islands with remote-sensing instruments and an undersea robot named Toby to explore a crustal anomaly in the Atlantic abyss.

The utopian impulse has migrated more widely. In the First Age its exemplar was an unknown island, Utopia, to the west, reported on by a Portuguese mariner. In the Second, it was a prelapsarian lost world, then a hidden valley, with a lamasery, Shangri-la, in the Himalayas stumbled upon by a Briton. In the Third, if Arthur C. Clarke is taken as oracle, it is an obelisk on the Saturnian moon, Iapetus, or if more popular versions have traction, a colony on Mars. The purpose of such a settlement seems to be morphing from status as a reincarnated promised land, a second-chance-better-than-Earth, to a refuge from an Earth trashed by its feckless inhabitants. The monastic vision is fully secularized, reduced to hamster-habitat huts on Martian sand.

What seems likely is that the magnificent bundling of cultural memes and practices that characterizes the Second Age will unravel. Exploration, commerce, colonization, science—all will find autonomous expressions in the new realms or form alliances of convenience for particular purposes. Commercial exploitation will stay in near-Earth orbit and near-continental shelves. Science will advance through new platforms and robots, gathering data. Colonization will remain a staple of fiction and astrofuturism. (The costly, ineffective folly of the space shuttle and ISS may be its death rattle. On the 50th anniversary of Apollo 8, former crewman Bill Anders publicly declared a crewed voyage to Mars absurd.) Conquest and conversion will be meaningless, unless they involve some sites where alien life exists, such that there is a nonhuman but still biotic encounter possible that will need protocols.[2]

To transform novelty into exploration requires a journey, preferably a trek beyond not just the quotidian but the unknown, maybe the unknowable. How much a civilization yearns or wishes for the quest will determine whether the activity will demand exploration, or be satisfied with science, adventure, and some simulacra of settlement. Alternatively, using exploration as a medium may make those other activities more attractive culturally. It may be possible to do more science—to understand the Kuiper belt, for example—with a Webb Space Telescope than with a spacecraft like *New Horizons*, but society may be more willing to support the science if it is packaged in a way that resonates with a valued past. Exploration may seem less a luxury accessory than a value added that makes a project succeed.

If the past is a guide, that will require a compelling rivalry. The classic formula involved competition among European nations or their neo-European settlement societies. The Cold War drove the early expansion of the Third Age and left a deposit of bases and missions that have continued, not unlike the settlements that Portugal and Spain left as berm after the great storm surge of the early First Age. Then other nations joined in emulation, or like the English and Dutch picked off those where the Iberians' reach had exceeded their grasp. Today the European Union has exerted itself mightily to absorb those ancient quarrels under a common purpose, which has also meant consolidating its exploration ambitions into a single cause or collaboration. The rise of the West has yielded to a rollback. Decolonialism has driven politics. Indigenous peoples are reclaiming ancestral lands, repatriating artifacts, and demanding parity with modern science for traditional ecological knowledge. The course of empire has passed to Asia.

It's possible that newcomers such as China, Japan, and India will enter—challenging and interloping as the Dutch and English did against the Iberians—and keep the pot boiling. More probably they will focus on commercial prospects in the near-environments of Earth and ocean, with a few bold bids elsewhere for prestige—not exploration as a cultural engine but expeditions as a marker of global status. For exploration as such they may see no purpose; after all they have no comparable tradition to that of the West. But without a serious rivalry it's hard to see how the Third Age can continue its early tempo. Without that recharge it may continue to run on ever diminishing power, like a far-ranging spacecraft slowly exhausting its internal power source.

After the first rush of discovery, a generation and a little more, a golden age passes. Its cultural energies dampen and exploration dims. Probably the Third Age is already in that second, silver age. The enterprise will likely continue to thin until at some point it ends.

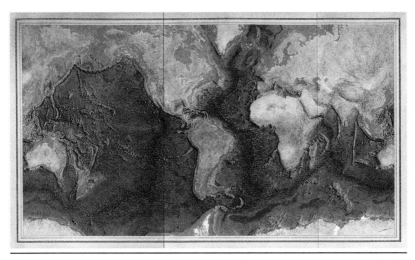

World seafloor map by Marie Tharp and Bruce Heezen, 1977—science as art further made artful through a painting by Heinrich Berann. Courtesy Library of Congress, Geography and Map Division.

Antarctic bedrock map, with East Antarctica left under its ice sheet. Courtesy NASA.

The Voyager twins and prizes of the Grand Tour. Montage with additions by Don Davis for JPL.

The DSV *Alvin*, in 1978, the year after the discovery of black smokers in the Pacific. Photo courtesy NOAA.

Curiosity on Mars, a selfie from the Vera Rubin Ridge (2018). Courtesy NASA/JPL-Caltec/MSSS.

Voyager 1 photo of the Earth and its Moon (1977). Courtesy JPL/
NASA.

Epilogue

Earth on the Edge

———————————

We were really after the picture of the Earth.[1]

—CANDICE HANSEN, VOYAGER IMAGING TEAM

Three are two big-screen prospects for the future of exploration, both of which address what might follow the Third Age.

One considers the Third Age as part of an ever-unfolding series of exploring eras. At some point the Third Age will yield to a Fourth. What such an age might constitute—what geographies it might trek to, with what cultural context, amid what moral drama—is difficult to imagine. It's easy to predict that the Third must end, it's impossible to forecast what might succeed it.

The other asks whether the end of the Third Age will mean the end of the Great Ages of Discovery. There was a time before that grand enterprise commenced, and surely there must be a time when it ceases, when the various motivating forces turn into something else or just turn off. The Great Ages are a cultural construct: they emerged at a particular time and place and society. There is no necessary reason why the project should continue indefinitely. But neither is there a reason to think it will end imminently.

What might an alternative look like? Assume that exploration is a response to humans' need to know bonded to the stimulus of adventure and that it will not be possible to shut those instincts off. Technology could continue to make them available. Space and ocean exploration have exhibited just such a hybrid of people and machines. Humans must rely on an artificial habitat; and spacecraft and rovers must possess a degree of autonomy. Designers have even come to speak of an "embodied experience" with regard to Mars rovers such that robots not only behave more like people, but that people identify with, and behave more in conformity with, the robots. Anthropologists have migrated from exploring for newfound tribes to studying explorers who are synchronized with machines on Mars.[2]

Pushed further this could lead to virtualized exploration. After all, curiosity can be satisfied by many means. Literature, art, laboratory science, invention, travel, collecting—individuals and societies have filled, and overtopped, their urge to pursue the curious (and avoid boredom) in many ways. Not all questing involves travel, not all travel requires exploring; not every curiosity bonds with adventuring, not all adventuring explores. Scholars have spent lifetimes pondering Shakespeare or pursuing Sophocles without exhausting their thirst to know or their excitement with the chase. Researches in field and lab have proved unbounded if not infinite. Geographic exploration is only one anodyne among an infinite set. Already both planetary and oceanographic discovery have apparatuses by which distant observation far removed from the operational staff, can share in discovery by watching in real time through the same streaming cameras. Geographic exploration is no longer something that an individual alone must do or experience; it can occur among millions simultaneously. It is only a short mutation of technology and a hop-skip of faith to have exploration venture into cyberspace altogether.

Engineers might take the process further. Science might find what physiological triggers exploration satisfies, and technology might discover virtual-reality games or designer drugs to satisfy them without the bother and boredom of hard travel, just as vitamin C can be distilled from oranges into a pill or amphetamines can bring a rush that in the past had to come from danger and adventure. The old quarrel between robots and astronauts might pale before the opportunity for virtual exploration or a telepresence that offers a comparable experience but with a real world rather than an imagined one. Players might experience the ecstasy of first-contact discovery without the annoyance of first-contact mosquitoes, fevers, troublesome natives, unreliable collaborators, and a nature reluctant to yield its secrets. There might be less cultural enthusiasm to sponsor expeditions with explorers as society's representatives if telepresence weakens the moral drama, or if everyone can experience the sensations for themselves.

Each of the motives bundled into exploration might be separately satisfied by other means. The reductionism that allows science to express itself in machines might equally be applied to disaggregate the experience of exploring and to address each urge independently of the others. It would be but a series of short steps to go from remote sensing to remote exploring to virtual exploring. *No Man's Sky*, an "action-adventure survival" game, positions players at the fringes of a galaxy containing 18,446,744,073,709,551,616 planets, each distinctive. NASA has websites devoted to cultivating a next generation. With Google it has developed a virtual exploration program for Mars and the Moon, and devised a spacewalk game for the International Space Station. Once again, this time in cyber form, an Ocean Sea has been populated with imagined isles awaiting discoverers. In brief, what distinguishes virtual, telepresent, or embedded from genuine exploration could blur, such that the impulses are triggered without the labor of trekking.[3]

Will such exercises consume the varied cultural urges that exploration has, for half a millennium, satisfied for Western civilization? Virtual reality is rolling over one frontier after another. A generation that has grown up immersed in high-tech gadgets may respond less to their ambient environment than to their digital one; their community is what they connect to electronically. Already nature enthusiasts are vocally worrying about the future of conservation as children come to experience nature not from personal contact but by digital simulacra. Some gamers prefer fantasy to real life. Will exploration be next? Could designer drugs trigger the ventral striatum to release neurostimulators in the absence of tangible stimuli? Could a tame exploration-designed LSD allow "trips" through space? Could cyberspace supply the wonder without the bother?

Maybe. Almost certainly the exploring tradition will adjust, perhaps like big game hunting morphing into nature tourism. Already millions of people can join Mars rovers through streaming video, a hybrid that keeps a traditional sense of exploration but redefines the explorer. The exploring impulse might veer into digital voyeurism, or become ceremonial like reenactments of Civil War battles or harden into a cultural legacy replayed by select elites like ballet and opera.

Or it may be that the crises on Earth subsumed under the concept of an Anthropocene become so undeniable, pressing, and dangerous that people decide to devote their energies to saving the world they live on rather than try to start over somewhere else. The globally connected world made possible by an exploring West threatens to terraform the one known-habitable planet into an unlivable one. When humanity, led by an industrial West, began to burn lithic landscapes as well as living ones, a process that started in rough synchronicity with the Second Age and went ballistic with the Third, its firepower became unbounded. It carried fire through deep time, extracting fuels from the geologic past and hurling their effluent into the geologic future. It displaced the Pleistocene's ice ages with a Pyrocene's fire age. It became a geologic force in itself. Reversing the myth of Prometheus, it has returned fire to the heavens.

The sense may grow that it is time to cease searching for new sources, and new worlds, and begin cleaning up the sinks left behind, the one world we may ever have. The urge for a better world will fold back on the home planet and make Old Earth a re-Newed World. The critical expedition is that of Spaceship Earth's voyage of survival. While the West's quest has encompassed the planet, and beyond, it has also brought Earth to the edge of habitability. Surely, the most enduring images from space travel are those of the Earth itself—the Earthrise of Apollo 8, the whole Earth of Apollo 17, the Earth and its Moon and the Earth in the solar system from *Voyager 1*.

The Great Ages of Discovery changed the course of Earth history. They were the means by which the peoples of Europe and their overseas settlements learned about a wider world, and then acted upon those findings, and so made their understanding global.

The Ages sparked a vast ecological reorganization of the planet that saw an immense mixing and sorting of plants, animals, diseases, and peoples on the scale of ice ages and shifting continents. They became a vector for industrialization, spread along routes of trade—themselves paths blazed by explorers. The industrialization of the planet has since propagated beyond the West's control and begin to unhinge the Earth's climate, midwife a mass extinction, and allow for melting ice and rising seas that are remaking continents and redefining the planet's character as an abode for life. Unlike many historical moments that saw empires come and go, peoples radiate and retreat, knowledge gained and lost, this one has been unique and irreversible. It could only happen once, and once it happened it could not be ignored or undone. The West left a very different Earth than it found.

The impact was felt not just by those peoples the West encountered, but by Western civilization itself. Discovery challenged text-based scholarship, helping wrench it into empirical science; challenged science, forcing it to absorb data and mutate into new disciplines; challenged art, urging it to frame new visions; challenged literature, recasting epics and redefining personal narratives; challenged politics, inspiring new institutions to deal with trade, rule, and colonization; challenged law, defining relationships among discoverers and between them and those they discovered; challenged religion, compelling Christendom to experience new moral universes, ultimately leading to notions of cultural relativism. Western civilization became something vastly different after the Great Ages than before.

Whether or not exploration survives as an institution, its history will endure as fact and story, as Western civilization's quest narrative. Like all foundational epics this one speaks to and magnifies the traits the culture most exhibits. Like great epics, too, it speaks not to simple triumphalism but equally to tragedy. It is all there, the best and the worst of the civilization, here expressed through a medium, a saga of exploration and discovery, distinctive to these peoples and their self-image.

The enterprise has made humanity more knowledgeable, if not wiser, and it has expanded the capacity of humans to inhabit Earth, without making their home planet more habitable. For now it continues, as the Third Age probes the future, perhaps poised for still greater revelations, perhaps to falter and fade, but this time it is not Europe that finds itself on the edge but the Earth.

Author's Note

This is an idea book.

The literature on exploration is vast. For Western civilization, it ranges across 600 years and the breadth of the solar system. And it is perennially popular. The classic tale has all that great literature requires—character, struggle, story. It builds on the archetypal quest narrative, modernizes the chivalric romance, and by fusing adventure with purposes larger than simple curiosity moves beyond the folkloric traveler's tale. For centuries exploring accounts have ranked among best sellers, and the most noteworthy bear continual retelling.

But this book is different. While it has characters, it is not character driven. While it involves events, it is not driven by plot. Rather it uses concepts to organize in a concise way the extraordinary outburst of exploration and geographic discovery conducted by the West since the 15th century. The organizing idea is a simple one: exploration is a cultural activity that displays the same kind of movement and rhythms as its sustaining society. Specifically, the grand saga of exploration divides into three long waves, or great ages of discovery, all of which share a common theme and continuities with the past but each of which expresses itself in a distinctive way. We are now living in the early phase of the third wave.

The idea of a great age of discovery is neither particularly old nor recent. In 1974, in a graduate seminar at the University of Texas-Austin, I listened as William Goetzmann read an essay in which he advanced the idea of a Second Great Age of Discovery, a change in exploration's style, dynamics, and purpose that began in the late 18th century, an idea he crystallized in *New Lands, New Men: America and the Second Great Age of Discovery* published in 1986. I quickly envisioned a Third Age that emerged in the post–World War II era and have used the great ages as an organizing device in a series of books—*Grove Karl Gilbert* (1980), *The Ice* (1986), *How the Canyon Became Grand* (1998), and *Voyager: Seeking Newer Worlds in the Third Great Age of Discovery* (2010). As the publication dates suggest, the great ages have not been the primary focus of my writing; that belongs with the history of fire, not the history of geographic discovery. Still, I continue to return to the concept as an organizing device and in *Voyager* I used the narrative of the mission as a sideways means to develop the notion of the three ages, especially the dimensions of the Third Age; the First Age figured only as a prelude. Inevitably, perhaps, as I return to the topic to consolidate my understanding, bits and passages from those books enter this one, like explorers carrying libraries of exploring books on their expeditions.

Throughout 45 years of intermittent thinking about the concept, the notion has grown and thickened, and I finally decided the time had come to devote a book to the concept itself, not use it as literary buttress and thematic rebar to strengthen other projects. *The Great Ages of Discovery* is the outcome. Any big-theme book is best treated either briefly or deeply. I decided it was best to go short. This is intended to lay out the basics of an idea, not serve as an encyclopedia of explorers and expeditions or a history sweeping all the sagas up into one massive narrative.

Any book so long in the making accrues many obligations; and one so long at sea as this one is encrusted with barnacles of debt.

Mine go back to that evening in Garrison Hall when I heard William Goetzmann read a short paper he was writing. Each of my subsequent books on exploration has elaborated on the theme of great ages, and each comes with its own baggage train of acknowledgments, parts of which also belong in this volume. Specifically, I would like to thank the Human Dimensions Faculty of the School of Life Sciences at Arizona State University for allowing me to break from my usual commitments and teach a course on the history of exploration, which forced me to think less allusively and more systematically about the subject. Most recently I want to thank Allyson Carter and the University of Arizona Press, which made one stretch to publish a series of fire books and is now making another to include exploration. Thanks, too, to Kerry Smith for his steady hand in copyediting and keeping the ship on the course I wanted it to go.

And of course, once again, it's a pleasure to thank Sonja for her encouragement, tolerance, and patience as I undertook yet another book. It seems I have a certain look about me when I begin plotting a new project. She recognizes it and finds quiet ways to help me see it into print.

Notes

PROLOGUE

1. Sauer, *Northern Mists*, 11.
2. Sauer, 4.
3. Edward L. Dreyer, *Zheng He: China and the Oceans in the Early Ming Dynasty, 1405–1433* (New York: Pearson Longman, 2007), quote from page 1.
4. Roger Crowley, *City of Fortune: How Venice Ruled the Seas* (New York: Random House, 2013), 135; Scammell, *World Encompassed*, 199.
5. Fernández-Armesto, *Times Atlas*, 43, 42.
6. "Medieval Sourcebook: Petrarch: The Ascent of Mount Ventoux," Fordham University, last modified January 2, 2020, https://sourcebooks.fordham.edu/source/petrarch-ventoux.asp.

THE RENAISSANCE EXPLORES

1. Quoted in Luc Cuyvers, *Into the Rising Sun: Vasco da Gama and the Search for the Sea Route to the East* (New York: TV Books, 1999), 84. Diffie and Winius, *Foundations*, 46–56, gives an informative and lively account of Ceuta.
2. Parry, *Establishment*, 11.
3. I found especially informative Fernández-Armesto's *Before Columbus*.

SAILS FOR THE WIND TO FILL

1. Luís Vaz de Camões, *The Lusíads*, trans. Landeg White (New York: Oxford University Press, 1997), canto 1, verse 27 (page 8).
2. One cannot go far in this era without encountering, again and again, but always with new information and fresh insights, the many works of J. H. Parry. He has

written in several books about the ships of discovery and their heritage. In this instance I have relied on his marvelous distillations in *Discovery of the Sea*, 3–24, and *Age of Reconnaissance*, 53–68.

3. Parry, *Age of Reconnaissance*; fleet composition from page 53, quote from page 54.

4. On the development of fighting ships, I have relied again on Parry, *Age of Reconnaissance*, 68, 114–127; and Parry, *Establishment*, 24–25.

5. Almeida quote from Diffie and Winius, *Foundations*, 229. The characterization of Morocco comes from pages 212–14. The book is a thorough and readable account of Portugal's empire of the east.

6. Parry, *Establishment*, 24–25.

7. Parry, *Discovery of the Sea*, 28.

8. Diffie and Winius, *Foundations*. On name of pilot, see Cuyvers, *Into the Rising Sun*, 98.

9. Parry, *Discovery of the Sea*, 41.

10. Parry, *Establishment*, 15.

11. Parry, *Discovery of the Sea*, 148–49.

12. Parry, *Discovery of the Sea*, 149; Parry, *Age of Reconnaissance*, 34.

GOD, GOLD, AND GLORY

1. Gomez Eannes de Azurara, *The Chronicle of the Discovery and Conquest of Guinea*, vol. 1, trans. Charles Raymond Beazley and Edgar Prestage (New York: Burt Franklin, 1963), 27.

2. Azurara, *Chronicle*, 28–30.

3. Azurara, *Chronicle*, 28, 33. Peter Martyr d'Anghiera, quoted in Daniel Boorstin, *The Discoverers* (New York: Random House, 1983), 145.

4. Columbus quoted in Ramon Iglesia, *Columbus, Cortes, and Other Essays*, trans. and ed. Lesley Byrd Simpson (Berkeley: University of California Press, 1969), 14. Bernal Díaz, *The Conquest of New Spain*, trans. J. M. Cohen (New York: Penguin Books, 1963), 15, 159, 274, 158.

5. Díaz, *Conquest*, 214.

6. Antonio Pigafetta, *Magellan's Voyage: A Narrative Account of the First Circumnavigation*, trans. and ed. R. A. Skelton (New York: Dover Publications, 1969), 37.

7. Díaz, *Conquest*, 214.

8. Díaz, 25.

9. de Camões, *Lusíads*, canto 4, verses 94–95.

10. Quotes and paraphrases from Pyne, *Voyager*, 129.

11. Fernão Mendes Pinto, *The Travels of Mendes Pinto*, ed. and trans. Rebecca D. Catz (Chicago: University of Chicago Press, 1989), 1.

12. Pinto, *Travels*, 4.
13. Azurara, *Chronicle*, 127.
14. Samuel Eliot Morison, interview quoted in *Arizona Republic*, May 16, 1976, 22.

WHERE NO HUMAN BEING EVER SAILED

1. On speculations about the Portuguese discovery of Australia, see Diffie and Winius, *Foundations*, which includes a master chronology of Portuguese expeditions and actions in the East, including a map that identifies Diogo Gomes de Sequeira on an expedition to New Guinea and northern Australia in 1525, but has no supporting text. A slightly obsessive but intriguing survey is offered in McIntyre, *Secret Discovery*.

ISLES

1. Much of the text in this section follows my description of islands in Pyne, *Voyager*, 180–85.
2. See Felipe Fernández-Armesto, *The Canary Islands After the Conquest: The Making of a Colonial Society in the Early Sixteenth Century* (Oxford: Clarendon Press, 1982), 205, fn. 6. On Columbus, Cecil Jane, trans. and ed., *The Four Voyages of Columbus* (New York: Dover Publications, 1988), ci. On Sancho Panza, see Fernández-Armesto, *Pathfinders*, 145.
3. Sidney Warhaft, ed., *Francis Bacon: A Selection of His Works* (Toronto: Macmillan of Canada, 1965), 447.

PORTUGUESE PARADIGM

1. On the strategic insights to the empire, see Diffie and Winius, *Foundations*, 227, and for a profile of Albuquerque, see Newitt, *History*, 81–90.
2. Boxer, *Portuguese*, 146–47.

THE ARMADA DE MOLUCCA CIRCUMNAVIGATES THE GLOBE

1. Pigafetta, *Magellan's Voyage*, 148.
2. Parts of my description of the "grand gesture" paraphrase or quote from Pyne, *Voyager*, 65.

ENCOUNTERING

1. Quoted in Bitterli, *Cultures*, 41.
2. On first contact, I follow closely, and quote extensively from, Pyne, *Voyager*, 258–62.

3. Crosby, *Ecological Imperialism*; Fernández-Armesto, *Before Columbus*.

4. Newitt, *History*, 45–46.

5. Newitt, 83–84.

6. Fernández-Armesto, *1492*, 78–81. Quotes from Newitt, *History*, 91–92.

7. Bob Connolly and Robin Anderson, *First Contact* (New York: Viking, 1987). The book accompanies a documentary film that uses footage from the Leahy expeditions.

8. Columbus quote from Morison, *Great Explorers*, 403. The two survivors' story from Díaz, *Conquest*, 59, 64–65. My account here closely follows or quotes from Pyne, *Voyager*, 262–64.

9. Donald M. Frame, trans., *The Complete Essays of Montaigne* (Stanford, Calif.: Stanford University Press, 1957), 158.

10. Frame, *Complete Essays*, 152–53, 158.

11. Frame, 158–59.

THE OTHER NEW WORLD

1. Álvaro Velho, *A Journal of the First Voyage of Vasco da Gama, 1497–1499*, Hakluyt Society, First Series, vol. 99, ed. E. G. Ravenstein (London: Hakluyt Society, 1898; repr., Farnham [England]: Ashgate, 2010), 48–49.

2. Columbus quote from Jane, *Four Voyages*, 10. Díaz quote from Díaz, *Conquest*, 86–87. Montaigne quote from Frame, *Complete Essays*, 159.

3. Adam Smith, *The Wealth of Nations*, books IV–V, 4th ed. (New York: Penguin Publishing Group, 2000), 209.

4. Martyr quoted in Iglesia, *Columbus*, 209.

5. Díaz, *Conquest*, 182–83.

6. Oviedo quote from Iglesia, *Columbus*, 216.

7. Diffie and Winius, *Foundations*, 181–82.

8. Since the 1980s the Canary Islands have reemerged as a template for European expansion. I have found the several works by Felipe Fernández-Armesto especially enlightening, and follow here his *Before Columbus*, 223–45; quote from 244.

9. For a good summary of Hernández, see Rick A. López, "Nature as Subject and Citizen in the Mexican Botanical Garden, 1787–1829," in *A Land Between Waters: Environmental Histories of Modern Mexico*, ed. Christopher R. Boyer (Tucson: University of Arizona Press, 2012), 73–99.

10. For the Iberians, see Daniela Bleichmar, ed., *Science in the Spanish and Portuguese Empires, 1500–1800* (Stanford, Calif.: Stanford University Press, 2009). Richard H. Grove includes a nice survey of extra-European gardens in his *Green Imperialism*.

11. On the Amadis allusion, see Díaz, *Conquest*, 214. For background see Irving A. Leonard, *Books of the Brave: Being an Account of Books and of Men in the Spanish Conquest and Settlement of the Sixteenth-Century New World* (Berkeley: University of California Press, 1992).

12. Source for data: Eltjo Buringh and Jan Luiten van Zanden, "Charting the 'Rise of the West': Manuscripts and Printed Books in Europe, a Long-Term Perspective from the Sixth Through Eighteenth Centuries," *Journal of Economic History* 69, no. 2 (2009): 409–45.

13. My treatment of England and especially Hakluyt paraphrases and quotes from my earlier text, Pyne, *Voyager*, 120–22.

14. John Parker, *Books to Build an Empire* (Amsterdam: N. Israel, 1965), 39.

15. Parker, *Books*, 102.

16. Richard Hakluyt, *Voyages and Discoveries* (New York: Penguin Classics, 1972), 31.

17. Hakluyt, 38.

18. Morison, *Great Explorers*, 127. See Grove, *Green Imperialism*, 249.

19. Lewis, *Muslim Discovery*, 296.

THE GREAT CONJUNCTION

1. Quote from Donald Weinstein, ed., *Ambassador from Venice: Pietro Pasqualigo in Lisbon, 1501* (Minneapolis: University of Minnesota Press, 1960), 46.

2. Crosby, *Ecological Imperialism*, 89.

3. John F. Richards, *The Unending Frontier: An Environmental History of the Early Modern World* (Berkeley: University of California Press, 2003).

4. Alfred W. Crosby, *The Columbian Exchange: Biological and Social Consequences of 1492* (Westport, Conn.: Greenwood Press, 1972).

5. Crosby, *Columbian Exchange*, is still the foundational work—the one that first organized these concepts around European contact and gave them a name. But see also William McNeill, *Plagues and Peoples* (New York: Knopf Doubleday, 1977), updated.

6. Jane, *Four Voyages*, 4, 5, 8, 14–16, 6, 12.

7. Bartolomé de Las Casas, *A Short Account of the Destruction of the Indies*, ed. and trans. Nigel Griffin (New York: Penguin Books, 1992).

THE DISCOVERERS AND THE DISCOVERED

1. Jane, *Four Voyages*, 2.

2. Quoted in William H. Goetzmann, *When the Eagle Screamed: The Romantic Horizon in American Diplomacy* (Norman: University of Oklahoma Press, 2006), 106.

3. Parry, *Age of Reconnaissance*, 313, 312.

4. Kris Lane, ed., *Defending the Conquest: Bernardo de Vargas Machuca's Defense and Discourse of the Western Conquests*, trans. Timothy F. Johnson (University Park: Pennsylvania State University Press, 2010), 16.
5. Jonathan Swift, *Gulliver's Travels* (New York: Barnes and Noble, 2004), 363.
6. de Camões, *Lusíads*, canto 4, verses 94–95.

EBB TIDE

1. Swift, *Gulliver's Travels*, 364.
2. Biographical information from John Knox Laughton, "Dampier, William," *Dictionary of National Biography, 1885–1900, Volume 14*, last modified December 29, 2012, https://en.wikisource.org/wiki/Dampier,_William_(DNB00).

CORPS OF DISCOVERY

1. Quoted in Douglas Botting, *Humboldt and the Cosmos* (New York: Harper and Row, 1973), 76.

THE ENLIGHTENMENT EXPLORES

1. Library of Congress, "Transcript: Jefferson's Instructions for Meriwether Lewis," *Rivers, Edens, Empires: Lewis & Clark and the Revealing of America*, accessed June 1, 2020, https://www.loc.gov/exhibits/lewisandclark/transcript57.html.
2. Humboldt quotes from Malcolm Nicolson, "Historical Introduction," in Alexander von Humboldt, *Personal Narrative of a Journey to the Equinoctial Regions of the New Continent*, trans. Jason Wilson (New York: Penguin Books, 1995), ix, xviii.
3. Library of Congress, "Transcript."

GRAND TOURS AND GREAT EXCURSIONS

1. Edward Smith, *Life of Sir Joseph Banks* (London, 1911), 15–16.
2. Quoted in Victor Wolfgang von Hagen, *South America Called Them: Explorations of the Great Naturalists: La Condamine, Humboldt, Darwin, Spruce* (New York: Alfred A. Knopf, 1945), 51. Mostly I rely on Larrie D. Ferreiro, *Measure of the Earth: The Enlightenment Expedition That Reshaped Our World* (New York: Basic Books, 2011), and as a popular account, Robert Whitaker, *The Mapmaker's Wife: A True Tale of Love, Murder, and Survival in the Amazon* (New York: Delta Trade Paperbacks, 2005).

3. I rely on Harry Woolfe, *The Transits of Venus* (Princeton, N.J.: Princeton University Press, 1959), and Andrea Wulf, *Chasing Venus: The Race to Measure the Heavens* (New York: Knopf, 2012).

4. Quote from Woolfe, *Transits of Venus*, 83.

5. Quoted in Hibbert, *Grand Tour*, 19.

6. Thomas Nugent, *The Grand Tour: A Journey Through the Netherlands, Germany, Italy, and France, Vol. 1* (London: D. Browne without Temple-Bar, A. Millar in the Strand, G. Hawkins in Fleetstreet, W. Johnston in St. Paul's Church-Yard, and P. Davey and B. Law in Ave-Mary-Lane, 1756), vii.

7. Nugent, *Grand Tour*, i–iii.

8. John Black, preface to *Political Essay on the Kingdom of New Spain*, vol. 1, by Alexander von Humboldt (New York: I. Riley, 1811), n.p.

9. James Boswell, *The Life of Samuel Johnson* (Garden City, N.Y.: Doubleday and Co., 1946), 355.

10. John Stuart Mill's reading list, accessed June 2018, https://www.laphamsquarterly .org/ways-learning/reading-list.

MOTIVES AND MOTIVATORS

1. Quote from Fergus Fleming, *Barrow's Boys* (New York: Grove Press, 1998), 1.

2. Goetzmann, *New Lands, New Men*, 178.

3. My account of Haënke follows closely and quotes Engstrand, *Spanish Scientists*, 46.

4. E. L. Godkin, "Was the Emin Pasha Expedition Piratical?," *Forum*, February 1891, 633, 645–46.

5. For a fascinating review of Thoreau's reading habits regarding exploration and travel, see Goetzmann, *New Lands, New Men*, 229–30.

SOMETHING OLD, SOMETHING NEW

1. Apsley Cherry-Garrard, *The Worst Journey in the World* (Harmondsworth, UK: Penguin Books, 1970), 643.

2. H. W. Menard, *Islands* (New York: Scientific American Books, 1986), 12–14.

3. See John Keay, *India Discovered: The Recovery of a Lost Civilization* (London: HarperCollins, 1981) and *The Great Arc: The Dramatic Tale of How India Was Mapped and Everest Named* (London: HarperCollins, 2000). For some color illustrations, see Fernández-Armesto, *Times Atlas*, 216–19.

4. Qianlong quote from Alain Peyrefitte, *The Immobile Empire*, trans. Jon Rothschild (New York: Alfred A. Knopf, 1992), xx.

ALEXANDER VON HUMBOLDT ASCENDS THE HEIGHTS

1. Quoted in de Terra, *Humboldt*, 320.
2. Alexander von Humboldt, *Personal Narrative*, trans. Jason Wilson (New York: Penguin Books, 1995), 7.
3. Andrea Wulf, ed., *Alexander Von Humboldt: Selected Writings* (New York: Everyman's Library, Alfred A. Knopf, 2018), 405.
4. Goetzmann, *New Lands, New Men*, 150–93.
5. "Preface," Alexander von Humboldt and Aimé Bonpland, *Personal Narrative of Travels to the Equinoctial Regions of the New Continent During the Years 1799–1804*, trans. Helen Maria Williams (Philadelphia, Pa.: M. Carey, 1815), iv.

CROSSING CONTINENTS

1. Bernard DeVoto, *The Journals of Lewis and Clark* (Boston: Houghton Mifflin, 1953), 279.
2. Ana Simões, Ana Carneiro, and Maria Paula Diogo, *Travels of Learning: A Geography of Science in Europe* (Boston: Kluwer Academic Publishers, 2003).
3. See Engstrand, *Spanish Scientists*, and Iris Enstrand, *The 1769 Transit of Venus Observed by Velasquez from Lower California*, Leaflet No. 419 (San Francisco, Calif.: Astronomical Society of the Pacific, 1964).
4. See Engstrand, *Spanish Scientists*.
5. William H. Goetzmann, "The Role of Discovery in American History," in *American Civilization*, ed. Daniel J. Boorstin (New York: McGraw-Hill, 1972), 25–36. Goetzmann's three volumes of American exploration are invaluable sources—*Army Exploration in the American West, 1802–1863* (New Haven, Conn.: Yale University Press, 1959); *Exploration and Empire*; and *New Lands, New Men*.
6. Quote from Donald Rayfield, *The Dream of Lhasa: The Life of Nikolay Przhevalsky (1839–88), Explorer of Central Asia* (Athens: Ohio University Press, 1976), 69.
7. Helen M. Rozwadowski, *Fathoming the Ocean: The Discovery and Exploration of the Deep Sea* (Cambridge, Mass.: Harvard University Press, 2005), 50.
8. Rozwadowski, *Fathoming the Ocean*, 27.
9. A concise summary of the voyage is found in Richard Corfield, *The Silent Landscape: The Scientific Voyage of HMS Challenger* (Washington, D.C.: Joseph Henry Press, 2003). For context within the oceanography of the day, see Rozwadowski, *Fathoming the Ocean*; and for a delightfully illustrated account, Eric Linklater, *The Voyage of the Challenger* (Garden City, N.Y.: Doubleday, 1972).
10. For a fascinating example of the endless search for Franklin, see Owen Beattie and John Geiger, *Frozen in Time: The Fate of the Franklin Expedition* (Toronto: Greystone Books, 1998).

SECOND LOOKS, REPEAT ENCOUNTERS

1. Clarence E. Dutton, *Tertiary History of the Grand Cañon District* (Washington, D.C.: U.S. Geological Survey, 1882; repr., Tucson: University of Arizona Press, 2001), 141.

2. On this improbable character, see John Glassie's aptly named *A Man of Misconceptions: The Life of an Eccentric in an Age of Change* (New York: Riverhead, 2012).

3. Lieutenant Joseph C. Ives, *Report upon the Colorado River of the West* (Washington, D.C.: Government Printing Office, 1861; repr., New York: Da Capo Press, 1969), 110.

4. Dutton, *Tertiary History*, 141, 150; Roosevelt quoted in Paul Schullery, ed., *The Grand Canyon: Early Impressions* (Boulder: Colorado Associated University Press, 1981), 101–2.

5. Dutton, *Tertiary History*, 143.

LOST HORIZONS

1. Jules Verne, *Seven Novels* (New York: Barnes and Noble, 2006), 539.

2. Apsley Cherry-Garrard, *The Worst Journey in the World* (New York: Penguin Books, 1970), 352, 356.

3. Roy Chapman Andrews, *This Business of Exploring* (New York: G. P. Putnam's Sons, 1935), xv.

4. Parts of this paragraph and the next derive from Stephen J. Pyne, *How the Canyon Became Grand: A Short History* (New York: Viking, 1998), 137–38. See also Harold Anthony, "The Facts About Shiva," 709–22, 766, and George Andrews, "Scaling Wotan's Throne," 723–24, 766, *Natural History* XL, no. 5 (December 1937).

MISSIONS OF DISCOVERY

1. Wilson, *IGY*, 105.

2. Arthur C. Clarke, *The Challenge of the Spaceship* (New York: Ballantine Books, 1961), 9–10.

MODERNISM EXPLORES

1. Wilson, *IGY*, 6.

2. This paragraph and the three that follow quote or paraphrase text from Pyne, *Voyager*, 48–49.

3. Wilson, *IGY*, 324.

THE GREAT GAME GOES GLOBAL, AND BEYOND

1. Wilson, *IGY*, 6.

2. This paragraph and the two preceding it are lightly rewritten versions of text that appears in Pyne, *Voyager*, 49.

3. Hamblin, *Oceanographers*, xviii.

4. Menard, *Ocean of Truth*, 38–40.

5. Peyrefitte, *Immobile Empire*, 546–47.

ICE

1. Richard Evelyn Byrd, *Little America: Aerial Exploration in the Antarctic, the Flight to the South Pole* (New York: G. P. Putnam's Sons, 1930), 342.

2. Wilson, *IGY*, 150, 288.

3. On the postwar transition (and the role of Byrd), see Lisle E. Rose, *Assault on Eternity: Richard E. Byrd and the Exploration of Antarctica, 1946–47* (Annapolis, Md.: Naval Institute Press, 1980).

4. Richard E. Byrd, *Alone* (New York: G. P. Putnam's Sons, 1938), 296. The oddness of the experience is enhanced in the book through the use of blue type.

SPACE

1. Carl Sagan, *Pale Blue Dot: A Vision of the Human Future in Space* (New York: Ballantine, 1994), 332.

2. Roger Launius, foreword, in Kraemer, *Beyond the Moon*, x. Murray quoted in David W. Swift, *Voyager Tales: Personal Views of the Grand Tour* (Reston, Va.: American Institute of Aeronautics and Astronautics, 1997), 214.

3. Sagan, *Pale Blue Dot*, vxii, 7.

ABYSS

1. Ballard, *Eternal Darkness*, 173.

2. Wilson, *IGY*, 245.

3. Wilson, *IGY*, 246. H. W. Menard, "Very Like a Spear," in Cecil J. Schneer, ed., *Two Hundred Years of Geology in America: Proceedings of the New Hampshire Bicentennial Conference on the History of Geology* (Hanover, N.H.: University Press of New England, 1979), 21–22.

4. Wilson, *IGY*, 220. Menard, "Very Like a Spear," 30.

5. Wilson, *IGY*, 252. For a popular introduction to Project Mohole, see Willard Bascom, *A Hole in the Bottom of the Sea: The Story of the Mohole Project* (Garden City, N.J.: Doubleday, 1961), which ends well before the Mohole limped to its conclusion. On Moondoggles, see Amitai Etzioni, *The Moon-Doggle. Domestic and International Implications of the Space Race* (Garden City, N.J.: Doubleday, 1964).

6. Ballard, *Eternal Darkness*, 127.

7. Ballard, 173.

8 Ballard, 311. Ballard's comment on the superiority of robots from an interview in Will Aslat et al., "The Next Frontier," in *The Universe Beneath the Sea* (Beverly Hills: World Almanac Video, 1999).

9. Quoted in Axel Madsen, *Cousteau: An Unauthorized Biography* (New York: Beaufort Books Publishers, 1986), 126.

10. Quote from Clarke from *The Challenge of the Sea* (New York: Dell, 1960), 161. This paragraph quotes or paraphrases from Pyne, *Voyager*, 284. Quote on Cousteau from Madsen, *Cousteau*, 227.

11. This paragraph quotes from or paraphrases Pyne, *Voyager*, 280.

12. The phrase became the title for Ballard's personal narrative, *Eternal Darkness*.

MODERN EXPLORATION, MODERNIST PARADOX

1. Amanda Kooser, "NASA Curiosity Rover Snaps Striking Mars Selfie Before Rolling Out," CNET, January 28, 2019, https://www.cnet.com/news/nasa-curiosity -rover-snaps-striking-mars-selfie-before-rolling-out/.

2. Byrd, *Little America*, 342.

3. For an interesting example, see William J. Clancey, *Working on Mars: Voyages of Scientific Discovery with the Mars Exploration Rovers* (Cambridge, Mass.: MIT Press, 2012).

4. Boies Penrose, *Travel and Discovery in the Renaissance 1420–1620* (New York: Atheneum, 1975), 118–19.

VOYAGER TRAVERSES THE SOLAR SYSTEM

1. Soderblum quote from JPL video, *And Then There Was Voyager*. I have confirmed his use of the quote and its original source by personal interview.

2. My thoughts on Voyager as a grand gesture derive from Pyne, *Voyager*, and in many places paraphrase or quote from Steve Pyne, "40 Years Ago NASA Launched Voyagers 1 and 2. It's Been a Stunning Ride," History News Network, September 1, 2017, http://historynewsnetwork.org/article/166802.

NEW REALMS, NEW REGIMES

1. Arvid Pardo, Agenda Item 92, U.N. General Assembly, Twenty-Second Session, *Official Records*, First Committee, 1515th Meeting, November 1, 1967, 12.

2. This paragraph quotes from and paraphrases from Pyne, *Voyager*, 333.

3. On the Antarctic Treaty, see U.S. Department of State, accessed June 2, 2020, https://2009-2017.state.gov/t/avc/trty/193967.htm. For overviews as the ATS has evolved, see Joyner, *Governing the Frozen Commons.*

4. U.S. Department of State, "Treaty on Principles Governing the Activities of States in the Exploration and Use of Outer Space, Including the Moon and Other Celestial Bodies," accessed June 2, 2020, https://2009-2017.state.gov/t/isn/5181.htm.

5. Pardo, Agenda Item 92, 2.

6. For the Law of the Sea Convention, and the American failure to sign, see U.S. State Department, "Law of the Sea Convention," accessed June 19, 2020, https://www.state.gov/law-of-the-sea-convention/. For a distilled summary of the many issues, see Hannigan, *Geopolitics of Deep Oceans.*

BEFORE AND AFTER

1. Robert Poole, *Earth*rise: *How Man First Saw the Earth* (New Haven, Conn.: Yale University Press, 2008), 199.

2. Poole, *Earth*rise, 199.

3. This paragraph quotes and paraphrases from Pyne, *Voyager*, 294.

4. Kuiper quote from G. P. Kuiper, ed., *The Earth as a Planet*, vol. 2, *The Solar System* (Chicago: University of Chicago Press, 1954), v. Baker quote from Robert H. Baker, *Astronomy*, 8th ed. (Princeton, N.J.: D. Van Nostrand Company), 310.

5. Parts of the paragraph quote Pyne, *Voyager*, 295.

6. Oran Nicks quote from Oran W. Nicks, *Far Travelers: The Exploring Machines*, NASA SP-480 (Washington, D.C.: National Aeronautics and Space Administration, 1985), 219.

7. Wilson quotes from, respectively, H. Takeuchi et al., *Debate About the Earth: Approach to Geophysics Through Analysis of Continental Drift* (San Francisco, Calif.: Freeman, Cooper, 1967), 244; Wilson, *IGY,* 190; and Takeuchi et al., *Debate,* 244–45.

8. Henry W. Menard, *Science: Growth and Change* (Cambridge, Mass.: Harvard University Press, 1971), 83.

9. Schlee, *Edge of an Unfamiliar World,* 19.

10. This paragraph quotes or paraphrases from Pyne, *Voyager*, 284.

11. This paragraph and the one that follows quotes or paraphrases from Pyne, *Voyager*, 292–93.

12. Rachel Carson, *The Sea Around Us* (New York: Oxford University Press, 1951), 216.

13. Wilson, *IGY,* 246.

14. Kurt Vonnegut Jr., *The Sirens of Saturn* (New York: Dell, 1959), 30.

LOOKING BACK, LOOKING AHEAD

1. Sagan, *Pale Blue Dot*, xvi.
2. On Anders quote, see Tom Green, "Sending Astronauts to Mars Would Be Stupid, Astronaut Says," BBC News, December 24, 2018, https://www.bbc.com/news/science-environment-46364179.

EPILOGUE

1. Candice J. (Candy) Hansen, in Swift, *Voyager Tales*, 324.
2. This paragraph and the four that follow quote from or paraphrase Pyne, *Voyager*, 306–8.
3. On *No Man's Sky*, see Raffi Khatchadourian, "World Without End," *New Yorker*, May 18, 2015, 48. A popular account of exploring by telepresence is Jeffrey Marlow, "Exploring the Oceans by Remote Control," *New Yorker*, May 28, 2019.

Selected Bibliography

SOURCES

Exploration is a topic equally blessed and burdened with sources—with atlases and maps, with original accounts and endless retellings, with biographies, with histories of expeditions, with surveys of countries and continents, across the Earth and beyond. *The Great Ages of Discovery* proposes a way to understand this abundance. It is not an encyclopedia of discovery, a biographical dictionary of explorers, an atlas of exploration history, or a panorama of archives. It is, instead, a conceptual frame by which to hold this cornucopia of information.

In what follows I propose a guide to some of the sources I found useful, emphasizing works of broad or interpretive sweep rather than amassing favorite accounts of individual explorers and expeditions. Where such works are used, they are cited in the endnotes.

GENERAL SURVEYS

Baker, J. N. L. *A History of Geographical Discovery and Exploration*. Rev. ed. New York: Cooper Square Publishers, 1967.

Beaglehole, J. C. *The Exploration of the Pacific*. 3rd ed. Stanford, Calif.: Stanford University Press, 1966.

Brown, Lloyd A. *The Story of Maps*. New York: Dover Publications, 1979.

Buisseret, David, ed. *The Oxford Companion to World Exploration*. Oxford: Oxford University Press, 2007.

Cannon, Michael. *The Exploration of Australia*. Sydney: Reader's Digest Services, 1987.

Crosby, Alfred W. *Ecological Imperialism: The Biological Expansion of Europe, 900–1900*. 2nd ed. Cambridge: Cambridge University Press, 2004.

Fernández-Armesto, Felipe. *Pathfinders: A Global History of Exploration*. New York: W. W. Norton, 2006.

Fernández-Armesto, Felipe, ed. *The Times Atlas of World Exploration: 3,000 Years of Exploring, Explorers, and Mapmaking*. New York: Times Books, 1991.

Goetzmann, William H. and Glyndwr Williams. *The Atlas of North American Exploration: From the Norse Voyages to the Race to the Pole*. New York: Prentice Hall, 1992.

Hakluyt Society, Series I and II. Some 290 volumes of original accounts in English translation. See also its periodical, *Journal of the Hakluyt Society*.

Mountfield, David. *A History of African Exploration*. Northbrook, Ill: Domus Books, 1976.

Mountfield, David. *A History of Polar Exploration*. New York: Dial Press, 1974.

Oxford University Press. *Oxford Atlas of Exploration*. 2nd ed. New York: Oxford University Press, 2008.

Reader's Digest. *Antarctica: Great Stories from the Frozen Continent*. Surrey Hills, NSW: Reader's Digest Services, 1985.

PRE-RENAISSANCE

Fernández-Armesto, Felipe. *Before Columbus: Exploration and Colonisation from the Mediterranean to the Atlantic, 1229–1492*. London: Macmillan Education, 1987.

Jones, Gwyn. *The Norse Atlantic Saga*. 2nd ed. New York: Oxford University Press, 1986.

Sauer, Carl O. *Northern Mists*. Berkeley: University of California Press, 1968.

Scammell, G. V. *The World Encompassed: The First European Maritime Empires c. 800–1650*. Berkeley: University of California Press, 1981.

FIRST AGE

The extraordinary sagas of the Great Voyages have attracted great historians. I've listed those books I found particularly enjoyable and enlightening.

Bitterli, Urs. *Cultures in Conflict: Encounters Between European and Non-European Cultures, 1492–1800*. Translated by Ritchie Robertson. Stanford, Calif.: Stanford University Press, 1989.

Fernández-Armesto, Felipe. *1492: The Year the World Began*. New York: HarperOne, 2009.

Fernández-Armesto, Felipe. *The Canary Islands After the Conquest: The Making of a Colonial Society in the Early Sixteenth Century*. Oxford: Clarendon Press, 1982.

Boxer, C. R. *The Dutch Seaborne Empire 1600–1800*. New York: Viking Penguin, 1965.

Boxer, C. R. *The Portuguese Seaborne Empire 1415–1825*. New York: Alfred A. Knopf, 1975.

DeVoto, Bernard. *The Course of Empire*. Boston: Houghton Mifflin, 1952.

Diffie, Bailey W., and George D. Winius. *Foundations of the Portuguese Empire 1415–1580*. Minneapolis: University of Minnesota Press, 1977.

Elliott, J. H. *Imperial Spain 1469–1716*. London: Penguin, 1963.

Lewis, Bernard. *The Muslim Discovery of Europe*. New York: W. W. Norton, 2001.

Mann, Charles C. *1493: Uncovering the New World Columbus Created*. New York: Alfred A. Knopf, 2011.

McIntyre, Kenneth Gordon. *The Secret Discovery of Australia: Portuguese Ventures 250 Years Before Captain Cook*. Sydney: Picador, 1977.

Morison, Samuel Eliot. *The Great Explorers: The European Discovery of America*. New York: Oxford University Press, 1978.

Newitt, Malyn. *A History of Portuguese Overseas Expansion, 1400–1668*. New York: Routledge, 2005.

Parry, J. H. *The Age of Reconnaissance: Discovery, Exploration and Settlement 1450–1650*. Berkeley: University of California Press, 1963.

Parry, J. H. *The Discovery of South America*. New York: Taplinger, 1979.

Parry, J. H. *The Discovery of the Sea*. Berkeley: University of California Press, 1974.

Parry, J. H. *The Establishment of the European Hegemony 1415–1715: Trade and Exploration in the Age of the Renaissance*. 3rd ed., rev. New York: Harper Torchbooks, 1966.

Parry, J. H. *The Spanish Seaborne Empire*. Berkeley: University of California Press, 1990.

Penrose, Boies. *Travel and Discovery in the Renaissance 1420–1640*. New York: Atheneum, 1975. The most distilled, comprehensive survey in text.

Sauer, Carl O. *The Early Spanish Main*. Berkeley: University of California Press, 1966.

Sauer, Carl O. *Seventeenth Century North America*. Berkeley: Turtle Island, 1980.

Sauer, Carl O. *Sixteenth Century North America*. Berkeley: University of California Press, 1971.

SECOND AGE

With the greatest number of expeditions and explorers, the Second Age has the grandest sweep of publications. I've selected from that vast number a mix of scholarly and popular books that go beyond individual stories.

Aldrich, Robert. *Greater France: A History of French Oversea Expansion*. New York: St. Martin's Press, 1996.

Alexander von Humboldt: Netzwerke des Wissens. Ostfildern: Hatje Cantz, 1999.

Becker, Peter. *The Pathfinders: The Saga of Exploration in Southern Africa*. Harmondsworth, UK: Penguin, 1985.

Blainey, Geoffrey. *A Land Half Won*. South Melbourne: Macmillan Company of Australia, 1980. Very readable survey of European contact with Australia.

Blunt, Wilfred. *Linnaeus: The Compleat Naturalist*. Princeton, N.J.: Princeton University Press, 2001.

de Terra, Helmut. *Humboldt: The Life and Times of Alexander von Humboldt, 1769–1859*. New York: Octagon Books, 1979.

Engstrand, Iris H. W. *Spanish Scientists in the New World: The Eighteenth-Century Expeditions*. Seattle: University of Washington Press, 1981.

Goetzmann, William H. *Exploration and Empire: The Explorer and the Scientist in the Winning of the American West*. New York: Alfred A. Knopf, 1967.

Goetzmann, William H. *New Lands, New Men: America and the Second Great Age of Discovery*. New York: Viking, 1986.

Grove, Richard H. *Green Imperialism: Colonial Expansion, Tropical Island Edens, and the Origins of Environmentalism, 1600–1860*. Cambridge: Cambridge University Press, 1995.

Hall, D. H. *History of the Earth Sciences During the Scientific and Industrial Revolutions, with Special Emphasis on the Physical Geosciences*. Amsterdam: Elsevier Scientific Publishing, 1976.

Hibbert, Christopher. *The Grand Tour*. New York: Hamlyn, 1974.

Jeal, Tim. *Stanley: The Impossible Life of Africa's Greatest Explorer*. New Haven, Conn.: Yale University Press, 2007.

Lloyd, T. O. *The British Empire 1558–1983*. New York: Oxford University Press, 1984.

Moorehead, Alan. *The Fatal Impact: The Invasion of the South Pacific 1767–1840*. New York: Harper and Row, 1987.

THIRD AGE

Most of the significant stories, apart from the most popular, are only recently told or still await telling. Scholarship lags, and when events are still unfolding, the vacuum fills with personal accounts and journalist reports. Below is a roster I found helpful.

Ballard, Robert, with Will Hively. *The Eternal Darkness: A Personal History of Deep-Sea Exploration*. Princeton, N.J.: Princeton University Press, 2000.

Belanger, Dian Olson. *Deep Freeze: The United States, the International Geophysical Year, and the Origins of Antarctica's Age of Science*. Boulder: University Press of Colorado, 2006.

Burrows, William E. *Exploring Space: Voyages in the Solar System and Beyond*. New York: Random House, 1990.

Burrows, William E. *This New Ocean: The Story of the First Space Age*. New York: Modern Library, 1999.

Chaikin, Andrew. *A Man on the Moon*. New York: Penguin, 1994.

Cooper, Henry S. F., Jr. *The Evening Star: Venus Observed*. Baltimore, Md.: Johns Hopkins University Press, 1994.

Cooper, Henry S. F., Jr. *Imaging Saturn: The Voyager Flights to Saturn*. New York: Holt, Rinehart and Winston, 1982.

Cooper, Henry S. F., Jr. *The Search for Life on Mars: Evolution of an Idea*. New York: Holt, Rinehart, and Winston, 1981.

Hamblin, Jacob Darwin. *Oceanographers and the Cold War: Disciples of Marine Science*. Seattle: University of Washington Press, 2005.

Hannigan, John. *The Geopolitics of Deep Oceans*. Cambridge: Polity Press, 2016.

Hartmann, William K., et al, eds. *In the Stream of Stars: The Soviet/American Space Art Book*. New York: Workman, 1990.

Harvey, Brian. *Russia in Space: The Failed Frontier?* Chicheser, UK: Praxis, 2001.

Harvey, Brian. *Russian Planetary Exploration: History, Development, Legacy and Prospects*. Chichester, UK: Praxis, 2007.

Hellwarth, Ben. *SEALAB: America's Forgotten Quest to Live and Work on the Ocean Floor*. New York: Simon and Schuster, 2012.

Hsü, Kenneth J. *The Mediterranean Was a Desert: A Voyage of the* Glomar Challenger. Princeton, N.J.: Princeton University Press, 1983.

Joyner, Christopher C. *Governing the Frozen Commons: The Antarctic Regime and Environmental Protection*. Columbia: University of South Carolina Press, 1998.

Kilgore, De Witt Douglas. *Astrofuturism: Science, Race, and Visions of Utopia in Space*. Philadelphia: University of Pennsylvania Press, 2003.

Koslow, Tony. *The Silent Deep: The Discovery, Ecology and Conservation of the Deep Sea*. Chicago: University of Chicago Press, 2007.

Kraemer, Robert S. *Beyond the Moon: A Golden Age of Planetary Exploration 1971–1978*. Washington, D.C.: Smithsonian Institution Press, 2000.

Kroll, Gary. *America's Ocean Wilderness: A Cultural History of Twentieth-Century Exploration*. Lawrence: University Press of Kansas, 2008.

Launius, Roger D. *Apollo's Legacy: Perspectives on the Moon Landings*. Washington, D.C.: Smithsonian Books, 2019.

Launius, Roger D., and Howard E. McCurdy. *Robots in Space: Technology, Evolution, and Interplanetary Travel*. Baltimore, Md.: Johns Hopkins Press, 2008.

McDougall, Walter A. *The Heavens and the Earth: A Political History of the Space Age.* New York: Basic Books, 1985.

Menard, H. W. *The Ocean of Truth: A Personal History of Plate Tectonics.* Princeton, N.J.: Princeton University Press, 1986.

Miller, Ron. *The Art of Space: The History of Space Art, from the Earliest Visions to the Graphics of the Modern Era.* Minneapolis, Minn.: Zenith Press, 2014.

Murray, Bruce. *Journey into Space: The First Thirty Years of Space Exploration.* New York: W. W. Norton, 1989.

Peterson, J. J. *Managing the Frozen South: The Creation and Evolution of the Antarctic Treaty System.* Berkeley: University of California Press, 1988.

Pyne, Stephen J. *Voyager: Seeking Newer Worlds in the Third Great Age of Discovery.* New York: Viking, 2010.

Roland, Alex. *A Spacefaring People: Perspectives on Early Spaceflight.* Washington, D.C.: NASA, 1985.

Schlee, Susan. *The Edge of an Unfamiliar World: A History of Oceanography.* New York: E. P. Dutton, 1973.

Sullivan, Walter. *Assault on the Unknown: The International Geophysical Year.* New York: McGraw-Hill, 1961.

Westwick, Peter J. *Into the Black: JPL and the American Space Program 1976–2004.* New Haven, Conn.: Yale University Press, 2007.

Wilkins, Don E. *To a Rocky Moon: A Geologist's History of Lunar Exploration.* Tucson: University of Arizona Press, 1993.

Wilson, J. Tuzo. *IGY: The Year of the New Moons.* New York: Alfred A. Knopf, 1961.

Index

About the Author

Stephen J. Pyne is an emeritus professor at Arizona State University. Among his exploration books are *The Ice*, *How the Canyon Became Grand*, and *Voyager*. Among recent fire books are *Between Two Fires: A Fire History of Contemporary America* and a suite of regional reconnaissances under the rubric To the Last Smoke. He lives in Queen Creek, Arizona.